Design Technologies for Green and Sustainable Computing Systems

Partha Pratim Pande • Amlan Ganguly
Krishnendu Chakrabarty

Editors

Design Technologies
for Green and Sustainable
Computing Systems

 Springer

Editors
Partha Pratim Pande
School of EECS
Washington State University
Pullman, WA, USA

Amlan Ganguly
Department of Computer Engineering
Rochester Institute of Technology
Rochester, NY, USA

Krishnendu Chakrabarty
ECE
Duke University
Durham, NC, USA

ISBN 978-1-4899-9641-1   ISBN 978-1-4614-4975-1 (eBook)
DOI 10.1007/978-1-4614-4975-1
Springer New York Heidelberg Dordrecht London

Preface

Modern large-scale computing systems, such as data centers and high-performance computing (HPC) clusters, are severely constrained by power and cooling costs for solving extreme-scale (or exascale) problems. The relentless increase in power consumption is of growing concern due to several reasons, e.g., cost, reliability, scalability, and environmental impact. A report from the Environmental Protection Agency (EPA) indicates that the nation's servers and data centers alone use about 1.5% of the total national energy consumed per year, at a cost of approximately $4.5 billion. The growing energy demands in data centers and HPC clusters are of utmost concern and there is a need to build efficient and sustainable computing environments that reduce the negative environmental impacts. Emerging technologies to support these computing systems are therefore of tremendous interest. Power management in data centers and HPC platforms is getting significant attention both from academia and industry. The power efficiency and sustainability aspects need to be addressed from various angles that include system design, computer architecture, programming language, compilers, networking, etc.

The aim of this book is to present several articles that highlight the state of the art on *Sustainable and Green Computing Systems*. While bridging the gap between various disciplines, this book highlights new sustainable and green computing paradigms and presents some of their features, advantages, disadvantages, and associated challenges. This book consists of nine chapters and features a range of application areas, from sustainable data centers, to run-time power management in multicore chips, green wireless sensor networks, energy efficiency of servers, cyber physical systems, and energy-adaptive computing. Instead of presenting a single, unified viewpoint, we have included in this book a diverse set of topics so that the readers have the benefit of variety of perspectives.

We hope that the book serves as a timely collection of new ideas and information to a wide range of readers from industry, academia, and national laboratories. The chapters in this book will be of interest to a large readership due to their interdisciplinary nature.

Washington State University, Pullman, USA   Partha Pratim Pande
Rochester Institute of Technology, Rochester, USA   Amlan Ganguly
Duke University, Durham, USA   Krishnendu Chakrabarty

Contents

Chapter 1
Fundamental Limits on Run-Time Power Management Algorithms for MPSoCs

Siddharth Garg, Diana Marculescu, and Radu Marculescu

1.1 Introduction

Enabled by technology scaling, information and communication technologies now constitute one of the fastest growing contributors to global energy consumption. While the energy per operation, joules per bit switch for example, goes down with technology scaling, the additional integration and functionality enabled by smaller transistors has resulted in a net growth in energy consumption. To contain this growth in energy consumption and enable sustainable computing, chip designers are increasingly resorting to run-time energy management techniques which ensure that each device only dissipates as much power as it needs to meet the performance requirements. In this context, MPSoCs implemented using the multiple Voltage Frequency Island (VFI) design style have been proposed as an effective solution to decrease on-chip power dissipation [10, 17]. As shown in Fig. 1.1a, each island in a VFI system is locally clocked and has an independent voltage supply, while inter-island communication is orchestrated via mixed-clock, mixed-voltage FIFOs. The opportunity for power savings arises from the fact that the voltage of each island can be independently tuned to minimize the system power dissipation, both dynamic and leakage, under performance constraints.

   In an ideal scenario, each VFI in a multiple VFI MPSoC can run at an arbitrary voltage and frequency so as to provide the lowest power consumption at the desired performance level. However, technology scaling imposes a number of fundamental constraints on the choice of voltage and frequency values, for example, the difference between the maximum and minimum supply voltage has

S. Garg (✉)
University of Waterloo, 200 Univ. Avenue W., Waterloo, ON, Canada
e-mail: siddharth.garg@uwaterloo.ca

D. Marculescu • R. Marculescu
Carnegie Mellon University, 5000 Forbes Ave., Pittsburgh, PA, USA
e-mail: dianam@ece.cmu.edu; radum@ece.cmu.edu

P.P. Pande et al. (eds.), *Design Technologies for Green and Sustainable Computing Systems*,
DOI 10.1007/978-1-4614-4975-1_1, © Springer Science+Business Media New York 2013

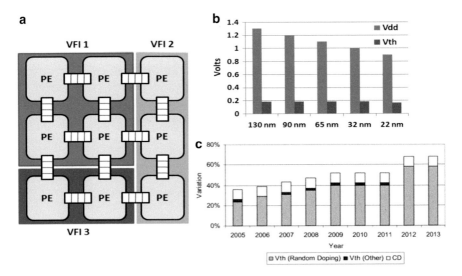

Fig. 1.1 (**a**) A multiple VFI system with three VFIs. (**b**) Decreasing difference between V_{dd} and V_{th} with technology scaling [27]. (**c**) Increasing process variation with technology scaling as outlined by the ITRS 2009

been shrinking with technology scaling which results in a reduced dynamic range to make DVFS decisions. While the problem of designing appropriate dynamic voltage and frequency scaling (DVFS) control algorithms for VFI systems has been addressed before by a number of authors [2, 16, 17, 25],[1] no attention has been given to analyzing the *fundamental limits* on the capabilities of DVFS controllers for multiple VFI systems.

Starting from these overarching ideas, we specifically focus on three technology driven constraints that we believe have the most impact on DVFS controller characteristics: (1) reliability-constrained upper-limits on the maximum voltage and frequency at which any VFI can operate; (2) inductive noise-driven limits on the maximum rate of change of voltage and frequency; and (3) the impact of manufacturing process variations. Figure 1.1b shows ITRS projections for supply voltage and threshold voltage scaling – assuming that the supply voltage range allowed during DVFS can swing between a fixed multiple of the threshold voltage and maximum supply voltage, it is clear that the available swing from minimum to maximum supply voltage is reducing. Similarly, Fig. 1.1c shows the increasing variations in manufacturing process variations with technology scaling, which eventually lead to significant core-to-core differences in power and performance characteristics on a chip. Finally, although not pictured in Fig. 1.1, in [15], the

---

[1]Note that an exhaustive list of prior work on DVFS control would be too lengthy for this manuscript. We therefore chose to detail only the publications that are most closely related to our work.

authors demonstrate the quadratic increase in peak voltage swing due to inductive noise (relative to the supply voltage) with technology scaling. Inductive noise is caused by sudden changes in the chip's power consumption and therefore DVFS algorithms must additionally be supply voltage noise aware.

Given the broad range of proposed DVFS control algorithms proposed in literature, we believe that it is insufficient to merely analyze the performance limits of a specific control strategy. The *only* assumption we make, which is common to a majority of the DVFS controllers proposed in literature, is that the goal of the control algorithm is to ensure that a reference state of the system is reached within a bounded number of control steps, for example, the occupancies of a pre-defined set of queues in the system are controlled to remain at pre-specified reference values. In other words, the proposed bounds are particularly applicable to DVFS control algorithms that, instead of directly minimizing total power dissipation (both static and dynamic), aim to do so indirectly by explicitly satisfying given performance/throughput constraints.

If the metric to be controlled is queue occupancy, we define the performance of a controller to be its ability to bring the queues, starting from an arbitrary initial state, back to their reference utilizations in a *desired, but fixed number of control intervals*. Given the technology constraints, our framework is then able to provide a *theoretical guarantee* on the existence of a controller that can meet this specification. The performance metric is a measure of the responsiveness of the controller to adapt to workload variations, and consequently reduce the power and energy dissipation when the workload demands do not require every VFI to run at full voltage and frequency.

1.2 Related Work and Novel Contributions

Power management of MPSoCs implemented using a multiple VFIs has been a subject to extensive research in the past, both from an *control algorithms* perspective and an *control implementation* perspective. Niyogi and Marculescu [16] presents an Lagrange optimization based approach to perform DVFS in multiple VFI systems, while in [25], the authors propose a PID DVFS controller to set the occupancies of the interface queues between the clock domains in a multiple clock-domain processor to reference values. In addition, [17] presents a state-space model of the queue occupancies in an MPSoC with multiple VFIs and proposes a formal linear feedback control algorithm to control the queues based on the state-space model. Carta [2] also uses a inter-VFI queue based formulation for DVFS control but makes use of non-linear feedback control techniques. However, compared to [17], the non-linear feedback control algorithm proposed by Carta et al. [2] can only be applied to simple pipelined MPSoC systems. We note that compared to [2, 17] and the other previous work, we focus on the fundamental limits of controllability of DVFS enabled multiple VFI systems. Furthermore, since we do not target a specific control algorithm, the results from our analysis are equally applicable to any of the control techniques proposed before. On a related note, feedback control techniques have recently been proposed for on-chip temperature management of multiple VFI systems [23,26], where, instead of queue occupancy, the goal is to keep the temperature of the system at or below a reference temperature. While outside the direct scope of this work, determining fundamental limits on the performance of on-chip temperature management approaches is an important avenue for future research.

Some researchers have recently discussed the practical aspects of implementing DVFS control on a chip, for example, tradeoffs between on-chip versus off-chip DC-DC converters [12], the number of discrete voltage levels allowed [5], and centralized versus distributed control techniques [1, 7, 18]. While these practical implementation issues also limit the performance of DVFS control algorithms, in this work we focus on more fundamental constraints mentioned before that arise from technology scaling and elucidate their impact on DVFS control performance from an algorithmic perspective.

Finally, a number of recent academic and industrial hardware prototypes have demonstrated the feasibility of enabling fine-grained control of voltage and frequency VFI-based multi-processor systems. These include the 167-core prototype designed by Truong et al. [22], the *Magali* chip [3], and the Intel 48-core *Single Chip Cloud (SCC)* chip [20] among others. The SCC chip, for example, consists of six VFIs with eight cores per VFI. Each VFI can support voltages between 0.7 and 1.3 V in increments of 0.1 V and frequency values between 100 and 800 MHz. This allows the chip's power envelope to be dynamically varied between 20 and 120 W.

As compared to the prior work on this topic, we make the following novel contributions:

- We propose a computationally efficient framework to analyze the impact of three major technology-driven constraints on the performance of DVFS controllers for multiple VFI MPSoCs.
- The proposed analysis framework *is not bound* to a specific control technique or algorithm. Starting from a formal state-space representation of the queues in an MPSoC, we provide theoretical bounds on the capabilities of *any* DVFS control technique; where we define the capability of a DVFS control algorithm to be its ability to bring the queue occupancies back to reference state starting from perturbed values.

We note that a part of this work, including figures, appeared in our prior publications [6,8].

1.3 Workload Control for VFI Based MPSoCs

The power management problem for VFI MPSoCs is motivated by the spatial and temporal workload variations observed in typical MPSoCs. In particular, to satisfy the performance requirements of an application executing on an MPSoC, it may not be required to run each core at full voltage and at its highest clock frequency,

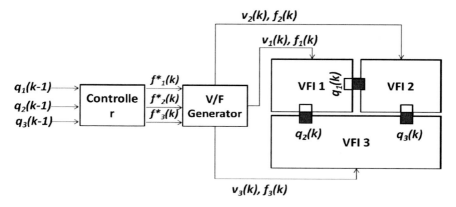

Fig. 1.2 Example of a VFI system with three islands and two queues

providing an opportunity to save power by running some cores at lower power and performance levels. In addition, looking at a specific core, its power and performance level may need to be changed temporally to guarantee that the performance specifications are met. In other words, the ideal DVFS algorithm for a multiple VFI MPSoC meets the performance requirements and simultaneously minimizes power dissipation (or energy consumption). While conceptually straightforward, it is not immediately clear how DVFS can be accomplished in real-time; towards this end, a number of authors have proposed queue stability based DVFS mechanisms. In essence, by ensuring that the queues in the system are neither too-full nor too-empty, it is possible to guarantee that the application demands are being met and, in addition, each core is running at the minimum speed required for it to meet these demands.

To mathematically describe queue-based DVFS control, we begin by briefly reviewing the state-space modeled developed in [17] to model the controlled queues in a multiple VFI system. We start with a design with N interface queues and M VFIs. An example of such a system is shown in Fig. 1.2, where $M = 3$ and $N = 2$. Furthermore, without any loss of generality, we assume that the system is controlled at discrete intervals of time, i.e., the kth control interval is the time period $[kT, (k + 1)T]$, where T is the length of a control interval.

The following notation can now be defined:

- The vector $Q(k) \in \mathbb{R}^N = [q_1(k), q_2(k), \ldots, q_N(k)]$ represents the vector of queue occupancies in the kth control interval.
- The vector $F(k) \in \mathbb{R}^M = [f_1(k), f_2(k), \ldots, f_M(k)]$ represents the frequencies at which each VFI is run in the kth control interval.
- λ_i and μ_i ($i \in [1, N]$) represent the average arrival and service rate of queue i, respectively. In other words, they represent the number of data tokens per unit of time a core writes to (reads from) the queue at its output (input). Due to workload variations, the instantaneous service and arrival rates will vary with time, for example, if a core spends more than average time in compute mode

on a particular piece of data, its read and write rates will drop. These workload dependent parameters can be obtained by simulating the system in the absence of DVFS, i.e., with each core running at full speed.

- The system matrix $B \in \mathbb{R}^{M \times N}$ is defined such that the (i, j)th entry of B is the rate of write (read) operations at the input (output) of the ith queue due to the activity in the jth VFI. We refer the reader to [17] for a detailed example on how to construct the system matrix.

The state-space equation that represents the queue dynamics can now simply be written as [17]:

$$Q(k + 1) = Q(k) + TBF(k) \tag{1.1}$$

The key observation is that, given the applied frequency vector $F(k)$ as a function of the control interval, this equation describes completely the evolution of queue occupancies in the system.

Also note that, as shown in Fig. 1.2, we also introduce an additional vector $F^*(k) = [f_1^*(k), f_2^*(k), \ldots, f_M^*(k)]$, which represents the *desired* control frequency values at control interval k. For a perfect system, $F^*(k) = F(k)$, i.e., the desired and applied control frequencies are the same. However, due to the technology driven constraints, the applied frequencies may deviate from the frequencies desired by the control, for example, if there is a limit on the maximum frequency at which a VFI can be operated. The technology driven deviations between the desired and actual frequency will be explained in greater detail in the next section.

1.4   Limits on DVFS Control

We now present the proposed framework to analyze the limits of performance of DVFS control strategies in the presence of technology driven constraints. To describe more specifically what we mean by performance, we define $Q_{ref} \in \mathbb{R}^N$ to be the desired *reference queue occupancies* that have been set by the designer. The reference queue occupancies represent the queue occupancy level at which the designer wants each queue to be stabilized; prior researchers have proposed workload characterization based techniques for setting the reference queue occupancies [25], but in this work we will assume that they are pre-specified. The proposed techniques, however, can be used to analyze any reference queue occupancy values selected by the designer or at run-time. We also assume that as a performance specification, the designer also sets a *limit*, J, that specifies the *maximum* number of control intervals that the control algorithm should take to bring the queues back from an arbitrary starting vector of queue occupancies, $Q(0)$, back to their reference occupancy values.[2] We expect that an appropriate choice of the specification, J, will be made by system-level designers, using, for example, transaction-level simulations, or even higher-level MATLAB or Simulink modeling methodologies.

Given this terminology, using Eq. 1.1, we can write the queue occupancies at the Jth control interval as [13]:

$$Q(J) = Q(0) + (TB) \sum_{k=0}^{J-1} F(k) \tag{1.2}$$

Since we want $Q(J) = Q_{ref}$, we can write:

$$(TB) \sum_{k=0}^{J-1} F(k) = (Q_{ref} - Q(0)) \tag{1.3}$$

1.4.1 Limits on Maximum Frequency

In a practical scenario, reliability concerns and peak thermal constraints impose an **upper limit** on the frequencies at which the VFIs can be clocked. As a result, if the desired frequency for any VFI is greater than its upper limit, the output of the VFI controller will saturate at its maximum value. For now, let us assume that each VFI in the system has a maximum frequency constraint $f_{MAX}^i (i \in [1, M])$. Therefore, we can write:

$$f_i(k) = min(f_{MAX}^i, f_i^*(k)) \quad \forall i \in [1, M] \tag{1.4}$$

Consequently, the system can be returned to its required state Q_{ref} in at most J steps *if and only if* the following system of linear equations has a feasible solution:

$$(TB) \sum_{k=0}^{J-1} F(k) = (Q_{ref} - Q(0)) \tag{1.5}$$

$$0 \leq f_i(k) \leq f_{MAX}^i \quad \forall k \in [0, J-1], \forall i \in [1, M] \tag{1.6}$$

Note that this technique only works for a *specific* initial vector of queue occupancies $Q(0)$; for example, $Q(0)$ may represent an initial condition in which all the queues in the system are full. However, we would like the system to be controllable in J time steps for a *set* of initial conditions, denoted by R_Q.

---

[2]The time index 0 for $Q(0)$ refers to a control interval at which the queue occupancies deviate from their steady-state reference values (Q_{ref}) due to changes in the workload behavior, and not necessarily to the time at which the system is started.

Let us assume that the set of initial conditions for which we want to ensure controllability is described as follows: $R_Q = \{Q(0) : A_Q Q(0) \leq B_Q\}$, where $A_Q \in \mathbb{R}^{P \times N}$ and $B_Q \in \mathbb{R}^P$ (P represents the number of linear equations used to describe R_Q). Clearly, the set R_Q represents a bounded *closed convex polyhedron* in $\mathbb{R}^N$. We will now show that to ensure controllability for all points in R_Q, it is sufficient to show controllability for each vertex of R_Q. In particular, without any loss of generality, we assume that R_Q has V vertices given by $\{Q^1(0), Q^2(0), \ldots, Q^V(0)\}$.

Lemma 1.1. *Any $Q(0) \in R_Q$ can be written as a convex combination of the vertices of R_Q, i.e., $\exists \{\alpha_1, \alpha_2 \ldots \alpha_V\} \in \mathbb{R}^N$ s.t. $\sum_{i=1}^{V} \alpha_i = 1$ and $Q(0) = \sum_{i=1}^{V} \alpha_i Q^i(0)$.*

Proof. The above lemma is a special case of the Krein-Milman theorem which states that a convex region can be described by the location of its corners or vertices. Please refer to [19] for further details.

Lemma 1.2. *The set of all $Q(0)$ for which Eqs. 1.5 and 1.6 admit a feasible solution is convex.*

Proof. Let $F^1(k)$ and $F^2(k)$ be feasible solutions for initial queue occupancies $Q^1(0)$ and $Q^2(0)$ respectively. We define $Q^3(0) = \alpha Q^1(0) + (1 - \alpha) Q^2(0)$, where $0 < \alpha < 1$. It is easily verified that $F^3(k) = \alpha F^1(k) + (1 - \alpha) F^2(k)$ is a feasible solution for Eqs. 1.5 and 1.6 with initial queue occupancy $Q^3(0)$.

Finally, based on Lemmas 1.1 and 1.2, we can show that:

Theorem 1.1. *Equations 1.5 and 1.6 have feasible solutions $\forall Q(0) \in R_Q$ if and only if they have feasible solutions $\forall Q(0) \in \{Q^1(0), Q^2(0), \ldots, Q^V(0)\}$.*

Proof. From Lemma 1.2 we know that any $Q(0) \in R_Q$ can be written as a convex combination of the vertices of R_Q. Furthermore, from Lemma 1.2, we know that, if there exists a feasible solution for each vertex in R_Q, then a feasible solution must exist for *any* initial queue occupancy vector that is a convex combination of the vertices of R_Q, which implies that a feasible solution must exist for any vector $Q(0) \in R_Q$.

Theorem 1.1 establishes necessary and sufficient conditions to *efficiently* verify the ability of a DVFS controller to bring the system back to its reference state, Q_{ref}, in J control intervals starting from a large set of initial states, R_Q, without having to independently verify that *each* initial state in R_Q can be brought back to the reference state. Instead, Theorem 1.1 proves that it is sufficient to verify the controllability for only the set of initial states that form the vertices of R_Q. Since the number of vertices of R_Q is obviously much smaller than the total number of initial states in R_Q, this significantly reduces the computational cost of the proposed framework.

In practice, the region of initial states R_Q will depend on the behavior of the workload, since queue occupancies that deviate from the reference values are observed due to changes in workload behavior away from the steady-state behavior, for example, a bursty read or a bursty write. While it is possible to obtain R_Q from extensive simulations of real workloads, R_Q can be defined conservatively as follows: $R_Q = \{Q(0) : 0 \le q_i(0) \le q_{MAX}^i\}, \forall i \in [1, N]$, where q_{MAX}^i is the physical queue length of the ith queue in the system. In other words, the conservative definition of R_Q implies a case in which, at any given point of time, each queue can have an occupancy between empty and full, irrespective of the other queues occupancies. In reality, the set R_Q can be much smaller, if for example, it is known that one queue is always full when the other is empty. Nonetheless, henceforth we will work with the conservative estimate of R_Q.

1.4.2 Inductive Noise Constraints

A major consideration for the design of systems that support dynamic voltage and frequency scaling is the resulting inductive noise (also referred to as the di/dt noise) in the power delivery network due to sudden changes in the power dissipation and current requirement of the system. While there exist various circuit-level solutions to the inductive noise problem, such as using large decoupling capacitors in the power delivery network or active noise suppression [11], it may be necessary to additionally constrain the maximum frequency increment from one control interval to another in order to obviate large changes in the power dissipation characteristics within a short period of time.

Inductive noise constraints can be modeled in the proposed framework as follows:

$$|f_i(k + 1) - f_i(k)| \le f_{step}^i \quad \forall i \in [1, M], \forall k \in [0, J - 1] \qquad (1.7)$$

where f_{step}^i is the maximum frequency increment allowed in the frequency of VFI i. Equation 1.7 can further be expanded as linear constraints as follows:

$$f_i(k + 1) - f_i(k) \le f_{step}^i \quad \forall i \in [1, M], \forall k \in [0, J - 1] \qquad (1.8)$$

$$-f_i(k + 1) + f_i(k) \le f_{step}^i \quad \forall i \in [1, M], \forall k \in [0, J - 1] \qquad (1.9)$$

Together with Eqs. 1.5 and 1.6, Eqs. 1.8 and 1.9 define a linear program that can be used to determine the existence of a time-optimal control strategy.

Finally, we note that for Theorem 1.1 to hold, we need to ensure that Lemma 1.2 is valid with the additional constraints introduced by Eq. 1.7. We show that this is indeed the case.

Lemma 1.3. *The set of all $Q(0)$ for which Eqs. 1.5, 1.6 and 1.7 admit a feasible solution is convex.*

Proof. As before, let $F^1(k)$ and $F^2(k)$ be a feasible solutions for an initial queue occupancies $Q^1(0)$ and $Q^2(0)$ respectively. In Lemma 1.2 we showed that $F^3(k) = \alpha F^1(k) + (1-\alpha)F^2(k)$ is a feasible solution for Eqs. 1.5 and 1.6 with initial queue occupancy $Q^3(0)$. The desired proof is complete, if we can show that $F^3(k)$ also satisfies Eq. 1.7, i.e.,

$$|f_i^3(k+1) - f_i^3(k)| \le f_{step}^i \quad \forall i \in [1, M], \forall k \in [0, J-1] \tag{1.10}$$

where, we know that:

$$|f_i^3(k+1) - f_i^3(k)|$$
$$= |\alpha(f_i^1(k+1) - f_i^1(k)) + (1-\alpha)(f_i^2(k+1) - f_i^2(k))| \tag{1.11}$$

Using the identity $|x + y| \le |x| + |y|$, we can write:

$$|f_i^3(k+1) - f_i^3(k)|$$
$$\le \alpha|(f_i^1(k+1) - f_i^1(k))| + (1-\alpha)|(f_i^2(k+1) - f_i^2(k))|$$
$$\le \alpha f_{step}^i + (1-\alpha) f_{step}^i = f_{step}^i \tag{1.12}$$

Therefore a feasible solution exists with initial queue occupancies $Q^3(0)$.

Lemma 1.3 ensures that Theorem 1.1 still remains valid after the inductive noise constraints given by Eq. 1.7 are added to the original set of linear constraints. Recall that Theorem 1.1 is essential to minimize the computational cost of the proposed method.

We note that there might be other factors besides inductive noise that constrain the maximum frequency increment. For example, experiments on the Intel SCC platform illustrate that the time to transition from one voltage and frequency pair to another is proportional to the magnitude of voltage change [4]. Thus, given a fixed time budget for voltage and frequency transitions, the maximum frequency (and voltage) increment becomes constrained. In fact, in their paper, the authors note that the large overhead of changing voltage and frequency values has a significant impact on the ability of the chip to quickly react to workload variations. Although further investigation is required, we suspect that this is, in fact, because of the fundamental limits of controllability given the slow voltage and frequency transitions.

1.4.3 Process Variation Impact

In the presence of process variations, the operating frequency of each VFI at the same supply voltage will differ even if they are the same by design. The maximum frequency of each island is therefore limited by the operating frequency at the maximum supply voltage allowed by the process. In other words, under the impact of process variations, we must think of f^i_{MAX} as *random variables*, not deterministic limits on the frequency at which each VFI can operate.

Since the maximum frequency bounds, f^i_{MAX}, must now be considered as random variables, the linear programming framework described in the previous sections will now have a certain probability of being feasible, i.e., there might exist values of f^i_{MAX} for which it is not possible to bring the system back to steady state within J control intervals. We will henceforth refer to the probability that a given instance of a multiple VFI system can be brought back to the reference queue occupancies in J time steps as the *probability of controllability* (PoC).

We use Monte Carlo simulations to estimate the PoC, i.e., in each Monte Carlo run, we obtain a sample of the maximum frequency for each VFI, f^i_{MAX}, and check for the feasibility of the linear program defined by Eqs. 1.5, 1.6, 1.8 and 1.9. Furthermore, we are able to exploit the specific structure of our problem to *speed up the Monte Carlo simulations*. In particular, we note that, if a given vector of upper bounds, $f^{i,1}_{MAX}(i \in [1, M])$, has a feasible solution, then another vector, $f^{i,1}_{MAX}(i \in [1, M])$, where $f^{i,2}_{MAX} \geq f^{i,1}_{MAX} \forall i \in [1, M]$ must also have a feasible solution. Therefore, we do not need to explicitly check for the feasibility of the upper bound $f^{i,2}_{MAX}$ by calling a linear programming solver, thereby saving significant computational effort. A similar argument is valid for the infeasible solutions and is not repeated here for brevity. As it will be seen from the experimental results, the proposed Monte Carlo method provides significant speed-up over a naive Monte Carlo implementation.

1.4.4 Explicit Energy Minimization

Until now, we have discussed DVFS control limits from a purely performance perspective – i.e., how quickly can a DVFS controller bring a system with queue occupancies that deviate from the reference values back to the reference state. However, since the ultimate goal of DVFS control is to save power under performance constraints, it is important to directly include energy minimization as an objective function in the mathematical formulation.[3] If E_{ik} denotes the energy dissipated by VFI i in control interval k, we can write the total energy dissipated by the system in the J control steps as:

$$E_{total} = \sum_{i=1}^{M} \sum_{k=1}^{J} E_{ik} = \sum_{i=1}^{M} \sum_{k=1}^{J} Pow_i(f_i(k))T \qquad (1.13)$$

---

[3]In this work, we concentrate only on the dynamic power dissipation, although leakage power can also be included.

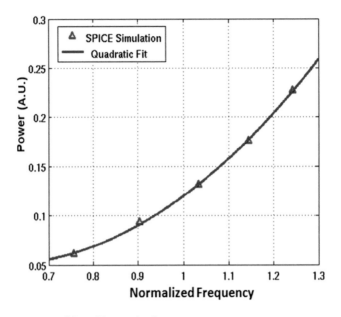

Fig. 1.3 *Power* versus f for a 90 nm technology

where $Pow_i(f_i(k))$ is the power dissipated by VFI i at a given frequency value. The mathematical relationship between the power and operating frequency can be obtained by fitting circuit simulation results at various operating conditions. Note that if only frequency scaling is used, the dynamic power dissipation is accurately modeled as proportional to the square of the operating frequency, but with DVFS (i.e., both voltage and frequency scaling), the relationship between frequency and power is more complicated and best determined using circuit simulations. Figure 1.3 shows SPICE simulated values for power versus frequency for a ring oscillator in a 90 nm technology node and the best quadratic fit to the SPICE data. The average error between the quadratic fit and the SPICE data is only 2%.

Along with the maximum frequency limit and the frequency step size constraints described before, minimizing E_{total} gives rise to a standard Quadratic Programming (QP) problem that can be solved efficiently to determine the control frequencies for each control interval that minimize total energy while bringing the system back to the reference state from an initial set of queue occupancies.

Using the quadratic approximation, we can write E_{total} as:

$$E_{total} = \sum_{i=1}^{M} \sum_{k=1}^{J} T(a_i f_i(k)^2 + b_i f_i(k) + c_i) \qquad (1.14)$$

where a_i, b_i and c_i are the coefficients obtained from the quadratic fit.

As in the case of time-optimal control, the energy minimization formulation provides an upper bounds on the maximum energy savings achievable by *any* DVFS control algorithm for a given set of parameters, i.e., an upper limit on the maximum frequency and frequency step size, the number of control intervals J and a vector of initial queue occupancies. Unfortunately, unlike the time-optimal control case, the bound on energy savings need to be computed for each possible vector of queue occupancies in R_Q, instead of just the vectors that lie on the vertices of R_Q.

Finally, we note that peak temperature is another important physical constraint in scaled technology nodes. Although we do not directly address peak temperature limits in this work, we note that the proposed formulation can potentially be extended to account for temperature constraints. If $Temp(k)$ and $Pow(k)$ are the vectors of temperature and power dissipation values for each VFI in the design, we can write the following state-space equation that governs the temperature dynamics:

$$Temp(k) = Temp(k-1) + \Theta Pow(k-1) \qquad (1.15)$$

where Θ accounts for the lateral flow of heat from one VFI to another. We have already shown that the power dissipation is a convex function of the operating frequency and the peak temperature constraint is easily formulated as follows:

$$Temp(k) \leq Temp_{max} \forall k \in [0, K-1] \qquad (1.16)$$

Based on this discussion, we conjecture that the peak temperature constraints are convex and can be efficiently integrated within the proposed framework.

1.5 Experimental Results

To validate the theory presented herein, we experiment on two benchmarks: (1) *MPEG*, is a distributed implementation of an MPEG-2 encoder with six ARM7-TDMI processors that are partitioned to form a three VFI system, as shown in Fig. 1.4a; and (2) *Star*, a five VFI system organized in a star topology as shown in Fig. 1.4b. The MPEG encoder benchmark was profiled on the cycle-accurate *Sunflower* MPSoC simulator [21] to obtain the average rates at which the VFIs read and write from the queues, as tabulated in Fig. 1.4a.[4] The arrival and service rates of the *Star* benchmark are randomly generated.

To begin, we first compute the nominal frequency values f_{NOM}^i of each VFI in the system, such that the queues remain stable for the nominal workload values. The maximum frequency constraint, f_{MAX}^i is then set using a parameter $\gamma = \frac{f_{MAX}^i}{f_{NOM}^i}$. In our experiments we use three values of $\gamma = \{1.1, 1.25, 1.5\}$, to investigate varying

---

[4]VFI 2 has the same read and write rates to its input and output queues, respectively.

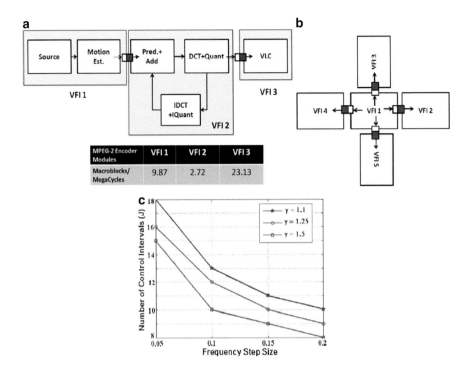

Fig. 1.4 (**a**) Topology and workload characteristics of the *MPEG* benchmark. (**b**) Topology of the *Star* benchmark. (**c**) Impact of γ and maximum frequency increment on the minimum number of control intervals, J

degrees of technology imposed constraints. Finally, we allow the inductive noise constrained maximum frequency increment to vary from 5 to 20% of the nominal frequency. We note that smaller values of *gamma* and of the frequency increment correlate with more scaled technology nodes, but we explicitly avoid annotating precise technology nodes with these parameters, since they tend to be foundry specific. For concreteness, we provide a case study comparing a 130 nm technology node with a 32 nm technology node using predictive technology models, later in this section.

Figure 1.4c shows the obtained results as γ and the maximum frequency step are varied for the *MPEG* benchmark. The results for *Star* benchmark are quantitatively similar, so we only show the graph for *MPEG* benchmark in Fig. 1.4c. As it can be seen, the frequency step size has a significant impact on the controllability of the system, in particular, for $\gamma = 1.5$ we see an 87% increase in the number of control intervals required to bring the system back to reference queue occupancies, J, while for $\gamma = 1.1$, J increases by up to 80%. The impact of γ itself is slightly more modest – we see a 20–25% increase in J as γ increases from 1.1 to 1.5.

To provide more insight in to the proposed theoretical framework, we plot in Fig. 1.5, the response of the time-optimal control strategy for the *MPEG* benchmark

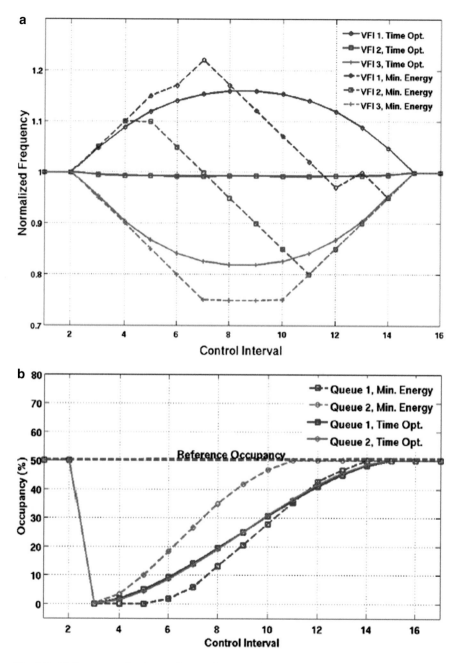

Fig. 1.5 (**a**) Response of a time-optimal and energy minimization controllers to deviation from the reference queue occupancies at control interval 2 for the *MPEG* benchmark. (**b**) Evolution of queue occupancies in the system with both queues starting from empty. Queue 1 is between VFI 1 and VFI 2, while Queue 2 is between VFI 2 and VFI 3

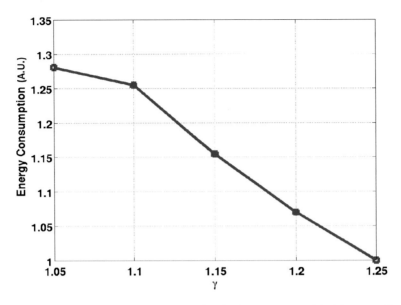

Fig. 1.6 Impact of γ on the energy savings achieved using an energy minimizing controller for the same performance specification J

when the queue occupancies of the two queues in the system drop to zero (i.e., both queues become empty) at control interval 2. As a result, the applied frequency values are modulated to bring the queues back to their reference occupancies within $J = 10$ control intervals. From Fig. 1.5a, we can clearly observe the impact of both the limit on the maximum frequency, and the limit on the maximum frequency increment, on the time-optimal control response. Figure 1.5b shows how the queue occupancies change in response to the applied control frequencies, starting from 0% occupancy till they reach their reference occupancies. From the figure we can clearly see that the controller with the energy minimization objective has a markedly different behaviour compared to the purely time-optimal controller, since, besides instead of trying to reach steady state as fast as possible, it tries to find the solution that minimized the energy consumption while approaching steady state. Numerically, we observe that the energy minimizing controller is able to provide up to 9% additional energy savings compared to the time-optimal controller for this particular scenario.

Figure 1.6 studies the impact of γ on the total energy required to bring the system back to steady state in a fixed number of control intervals assuming that the energy minimizing controller is used. Again, we can notice the strong impact of the ratio between the nominal and maximum frequency on the performance of the DVFS control algorithm – as γ decreases with technology scaling, Fig. 1.6 indicates that the energy consumed by the control algorithm will increase. This may seem counterintuitive at first, since lower γ indicates lower maximum frequency (for the same nominal frequency). However, note that any DVFS control solution that is feasible for a lower value of γ is also feasible for a higher γ value, while the converse is not true. In other words, the tighter *constraints* imposed by technology scaling reduce the energy efficiency of DVFS control.

Next, we investigate the impact of process variations on the probability of controllability (PoC), as defined in Sect. 1.4.3, of DVFS enabled multiple VFI systems. As mentioned before, because of process variations, the maximum frequency limits, f_{MAX}^i, are not fixed numbers, but random variables. For this experiment, we model the maximum frequency of each VFI as an independent normal distribution [14], and increase the standard deviation (σ) of the distribution from 2 to 10% of the maximum frequency. Finally, we use 5,000 runs of both naive Monte Carlo simulations and the proposed efficient Monte Carlo simulations (see Sect. 1.4.3) to obtain the PoC for various values of σ and for both benchmarks. From Fig. 1.7a, we can see that the proposed efficient version of Monte Carlo provides **significant speed-up** over the naive Monte Carlo implementation – on average, a 9× speed-up for the *MPEG* benchmark and a 5.6× speed-up for the *Star* benchmark – without any loss in accuracy.

From the estimated PoC values in Fig. 1.7b, we can see that the PoC of both *MPEG* and *Star* benchmarks are significantly impacted by process variations, though *MPEG* sees a greater degradation in the PoC, decreasing from 92% for $\sigma = 2\%$ to only 40% for $\sigma = 10\%$. On the other hand, the PoC of *Star* drops from 95 to 62% for the same values of σ. We believe that PoC of *Star* is hurt less by increasing process variations (as compared to *MPEG*) because for the *Star* benchmark, the PoC depends primarily on the maximum frequency constraint of only the central VFI (VFI 1), while for *MPEG*, all the VFIs tend to contribute to PoC equally. To explain the significance of these results, we point out that a PoC of 40% implies that, on average, 60% of the fabricated circuits will *not* be able to meet the DVFS control performance specification, irrespective of the control algorithm that is used. Of note, while the specific parameters used in the Monte Carlo simulations (for example, the value of γ at various technology nodes) are implementation dependent and may cause small changes in the PoC estimates in Fig. 1.7, the fundamental predictive nature of this plot will remain the same. This reveals the true importance of the proposed framework.

1.5.1 Case Study: 130 nm Versus 32 nm

While the experimental results shown so far have used representative numbers for the technology constraint parameters, it is instructive to examine how the proposed methodology can be used to compare two specific technology nodes. For this study, we compare an older 130 nm technology with a more current 32 nm technology node. For both cases, the technology libraries and parameters are taken from the publicly available PTM data [27]. In particular, the maximum supply voltage for the 130 nm technology is 1.3 V, while that for the 32 nm technology, it is only 0.9 V. On the other hand, to guarantee stability of SRAM cells, the minimum supply voltage

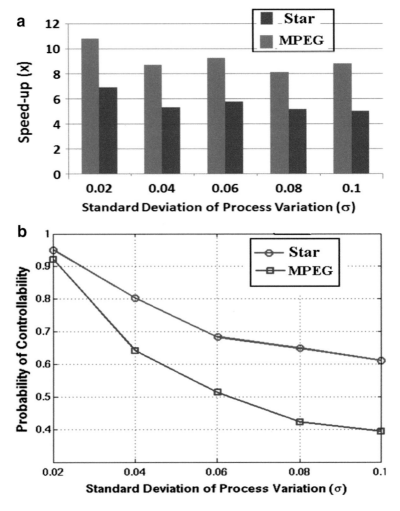

Fig. 1.7 (a) Speed-up (×) of the proposed efficient Monte Carlo technique to compute PoC compared to a naive Monte Carlo implementation. (b) PoC as a function of increasing process parameter variations for the *MPEG* and *Star* benchmarks

is limited by the threshold voltage of a technology node and is a fixed multiple of the threshold voltage. The threshold voltage for the two technologies is 0.18 and 0.16 V, respectively and the minimum voltage for each technology node is set at $4X$ its threshold voltage. It is clear that while the voltage in a 32 nm can only swing between 0.64 V → 0.9 V, for a 130 nm technology, the range is 0.72 V → 1.3 V.

To convert the minimum and maximum voltage constraints to constraints on the operating frequency, we ran SPICE simulations on ring oscillators (RO) constructed using two input NAND gates for both technology nodes at both operating points. ROs were chosen since they are commonly used for characterizing technology nodes, and to ensure that the results are not biased by any specific circuit implementation. Furthermore, although the quantitative results might be slightly different if a large circuit benchmark is used instead of an RO, we believe that the qualitative conclusions would remain the same. The maximum frequency for the 32 nm technology is 38% higher than its minimum frequency, while the maximum frequency for the 130 nm technology is 98% higher. This illustrates clearly the reduced range available to DVFS controllers in scaled technologies. Finally, assuming that the nominal frequency for both technology nodes is centered in its respective operating range, we obtain values of $\gamma_{32\,nm} = 1.159$ and $\gamma_{130\,nm} = 1.328$. For these constrains, and optimistically assuming that the inductive noise constraints do not become more stringent from one technology node to another, the number of control intervals required to bring the system back to steady state increases from 8 to 9 when going to a 32 nm technology. In addition, for a control specification of 9 control steps, the yield for a 130 nm design is 96% while the 32 nm design yields only 37%. Again, this is under the optimistic assumption that process variation magnitude does not increase with shrinking feature sizes – realistically, the yield loss for the 32 nm technology would be even greater.

We note that although our experimental results indicate that conventional DVFS techniques may become less effective with technology scaling due to the shrinking V_{dd} and V_{th} gap, and due to noise and variabilty effects, we view this as a challenge and not an insurmountable barrier. For example, alternative SRAM architectures have recently been proposed that enable potential scaling of $V_{DD.min}$ closer to or beyond the threshold voltage [24]. In addition, with increasing integration density, a case can be made for having a large number of heterogeneous cores on a chip and enabling only a subset of cores [9] at any given time, based on application requirements.

In fact, the increasing number of cores on a chip provides a greater spatial range and granularity of power consumption. If we look at a few technologies of interest, we can see that, while for 45 nm Intels SCC chip the number of cores is 48, under the same core power budget, at 32 and 22 nm, a chip will likely consist of 100 and 300 cores, respectively. Therefore, even if we conservatively (and unrealistically) assume that there are no opportunities for dynamic voltage scaling, a 300× spread in power consumption can be achieved for a 300 core system by turning an appropriate number of cores on or off. We believe that next generation DVFS algorithms will be accompanied by synergistic dynamic core count scaling algorithms to full exploit the available on-chip resources in the most power efficient way.

It is important to interpret the results in the correct context. In particular, our main claim in this paper is *not* that the baseline system performance and energy efficiency reduces with technology scaling, but that the performance of DVFS control algorithms, in terms of their ability to exploit workload variations, is expected to diminish in future technology nodes. At the same time, technology scaling also offers numerous *opportunities* to overcome the potential loss in DVFS control performance by, for example, allowing for an increased granularity of VFI partitioning and enabling more complex on-chip DVFS controllers that approach the theoretical performance limits. As such, we view our results not as negative results, instead as motivating the need for further research into overcoming the barriers imposed by technology scaling on fine-grained DVFS control.

1.6 Conclusion

We presented a theoretical framework to efficiently obtain the limits on the controllability and performance of DVFS controllers for multiple VFI based MPSoCs. Using a computationally efficient implementation of the framework, we present results, using both real and synthetic benchmarks, that explore the impact of three major technology driven factors – temperature and reliability constraints, maximum inductive noise constraints and process variations – on the performance bounds of DVFS control strategies. Our experiments demonstrate the importance of considering the impact of these three factors on DVFS controller performance, particularly since all three factors are becoming increasingly important with technology scaling.

Acknowledgements Siddharth Garg acknowledges financial support from the Conseil de Recherches en Sciences Naturelles et en Genie du Canada (CRSNG) Discovery Grants program. Diana Marculescu and Radu Marculescu acknowledge partial support by the National Science Foundation (NSF) under grants CCF-0916752 and CNS-1128624.

References

1. Beigne E, Clermidy F, Lhermet H, Miermont S, Thonnart Y, Tran XT, Valentian A, Varreau D, Vivet P, Popon X et al (2009) An asynchronous power aware and adaptive NoC based circuit. IEEE J Solid-State Circuit 44(4):1167–1177
2. Carta S, Alimonda A, Pisano A, Acquaviva A, Benini L (2007) A control theoretic approach to energy-efficient pipelined computation in MPSoCs. ACM Trans Embedded Comput Syst (TECS) 6(4):27–es
3. Clermidy F, Bernard C, Lemaire R, Martin J, Miro-Panades I, Thonnart Y, Vivet P, Wehn N (2010) MAGALI: a network-on-chip based multi-core system-on-chip for MIMO 4G SDR. In: Proceedings of the IEEE international conference on IC design and technology (ICICDT). Grenoble, France IEEE, pp 74–77
4. David R, Bogdan B, Marculescu R (2012) Dynamic power management for multi-cores: case study using the Intel SCC. In: Proceedings of VLSI SOC conference. Santa Cruz, CA IEEE
5. Dighe S, Vangal S, Aseron P, Kumar S, Jacob T, Bowman K, Howard J, Tschanz J, Erraguntla V, Borkar N et al (2010) Within-die variation-aware dynamic-voltage-frequency scaling core mapping and thread hopping for an 80-core processor. In: IEEE solid-state circuits conference digest of technical papers. San Francisco, CA IEEE, pp 174–175
6. Garg S, Marculescu D, Marculescu R, Ogras U (2009) Technology-driven limits on DVFS controllability of multiple voltage-frequency island designs: a system-level perspective. In: Proceedings of the 46th IEEE/ACM design automation conference, San Francisco, CA IEEE, pp 818–821
7. Garg S, Marculescu D, Marculescu R (2010) Custom feedback control: enabling truly scalable on-chip power management for MPSoCs. In: Proceedings of the 16th ACM/IEEE international symposium on low power electronics and design. Austin, TX ACM, pp 425–430

8. Garg S, Marculescu D, Marculescu R (2012) Technology-driven limits on runtime power management algorithms for multiprocessor systems-on-chip. ACM J Emerg Technol Comput Syst (JETC) 8(4):28

9. Goulding-Hotta N, Sampson J, Venkatesh G, Garcia S, Auricchio J, Huang PC, Arora M, Nath S, Bhatt V, Babb J et al (2011) The GreenDroid mobile application processor: an architecture for silicon's dark future. IEEE Micro 31:86–95

10. Jang W, Ding D, Pan DZ (2008) A voltage-frequency island aware energy optimization framework for networks-on-chip. In: Proceedings of the IEEE/ACM international conference on computer-aided design, San Jose

11. Keskin G, Li X, Pileggi L (2006) Active on-die suppression of power supply noise. In: Proceedings of the IEEE custom integrated circuits conference, San Jose. IEEE, pp 813–816

12. Kim W, Gupta MS, Wei GY, Brooks D (2008) System level analysis of fast, per-core DVFS using on-chip switching regulators. In: Proceedings of the 14th international symposium on high performance computer architecture. Salt Lake City, UT IEEE, pp 123–134

13. Kuo BC (1992) Digital control systems. Oxford University Press, New York

14. Marculescu D, Garg S (2006) System-level process-driven variability analysis for single and multiple voltage-frequency island systems. In: Proceedings of the 2006 IEEE/ACM international conference on computer-aided design. San Jose, CA

15. Mezhiba AV, Friedman EG (2004) Scaling trends of on-chip power distribution noise. IEEE Trans Very Large Scale Integr (VLSI) Syst 12(4):386–394

16. Niyogi K, Marculescu D (2005) Speed and voltage selection for GALS systems based on voltage/frequency islands. In: Proceedings of the 2005 conference on Asia South Pacific design automation, Shanghai

17. Ogras UY, Marculescu R, Marculescu D (2008) Variation-adaptive feedback control for networks-on-chip with multiple clock domains. In: Proceedings of the 45th annual conference on design automation. Annaheim, CA

18. Ravishankar C, Ananthanarayanan S, Garg S, Kennings A (2012) Analysis and evaluation of greedy thread swapping based dynamic power management for MPSoC platforms. In: Proceedings of the 13th international symposium on quality electronic design (ISQED), Santa Clara. IEEE, pp 617–624

19. Royden HL (1968) Real analysis. Macmillan, New York

20. Salihundam P, Jain S, Jacob T, Kumar S, Erraguntla V, Hoskote Y, Vangal S, Ruhl G, Borkar N (2011) A 2 Tb/s 6 × 4 mesh network for a single-chip cloud computer with DVFS in 45 nm CMOS. IEEE J Solid State Circuit 46(4):757–766

21. Stanley-Marbell P, Marculescu D (2007) Sunflower: full-system, embedded microarchitecture evaluation. In: De Bosschere K, Kaeli D, Stenström P, Whalley D, Ungerer T (eds) High performance embedded architectures and compilers. Springer, Berlin, pp 168–182

22. Truong D, Cheng W, Mohsenin T, Yu Z, Jacobson T, Landge G, Meeuwsen M, Watnik C, Mejia P, Tran A et al (2008) A 167-processor 65 nm computational platform with per-processor dynamic supply voltage and dynamic clock frequency scaling. In: Proceedings of the IEEE symposium on VLSI circuits. Honolulu, Hawaii IEEE, pp 22–23

23. Wang Y, Ma K, Wang X (2009) Temperature-constrained power control for chip multiprocessors with online model estimation. In: ACM SIGARCH computer architecture news, vol 37. ACM, pp 314–324

24. Wilkerson C, Gao H, Alameldeen AR, Chishti Z, Khellah M, Lu SL (2008) Trading off cache capacity for reliability to enable low voltage operation. In: ACM SIGARCH computer architecture news, vol 36. IEEE Computer Society, pp 203–214

25. Wu Q, Juang P, Martonosi M, Clark DW (2004) Formal online methods for voltage/frequency control in multiple clock domain microprocessors. ACM SIGOPS Oper Syst Rev 38(5): 248–259

26. Zanini F, Atienza D, Benini L, De Micheli G (2009) Multicore thermal management with model predictive control. In: Proceedings of the European conference on circuit theory and design. Antalya, Turkey IEEE, pp 711–714

27. Zhao W, Cao Y (2006) New generation of predictive technology model for sub-45 nm design exploration. In: Proceedings of ISQED, San Jose

Chapter 2
Reliable Networks-on-Chip Design
for Sustainable Computing Systems

Paul Ampadu, Qiaoyan Yu, and Bo Fu

2.1 Introduction

Since the 1 gigaFLOPS Cray 2 [1] was created in 1985, supercomputers have
shown a sustained growth [2]. Recently, the Chinese supercomputer, Tianhe-
1, achieves 2.5 petaFLOPS performance [3]. To further enhance the computing
capability, many-core systems have been attractive [4–6]. As bus-based interconnect
cannot keep pace with the increasing scale of many-core systems, networks-on-
chip (NoCs) have emerged as a promising on-chip interconnect infrastructure, in
terms of scalability, reusability, power efficiency and high throughput [7–9]. NoCs
have also demonstrated potential to manage the increasing complexity of on-chip
communication, by routing message packets over multi-hop network. As shown in
Fig. 2.1, a NoC is typically composed of, on-chip interconnect (i.e., link), routers,
and network interfaces.

Thanks to advanced technologies, the many-core systems consist of a large
number of computation and storage cores that operate at low-voltage levels,
attempting to break the power wall [2, 10, 11]. Unfortunately, the side effect is
the new reliability challenge. In fact, reliability becomes one of the most critical
challenges caused by technology scaling and increasing chip densities [12–15].
Nanometer fabrication processes inevitably result in defective components, which
lead to permanent errors [16]. As the critical charge of a capacitive node decreases

P. Ampadu (✉)
University of Rochester, Rochester, NY 14627, USA
e-mail: paul.ampadu@rochester.edu

Q. Yu
University of New Hampshire, Durham, NH 03824, USA
e-mail: qiaoyan.yu@unh.edu

B. Fu
Marvell Technology Group Ltd., Santa Clara, CA, USA
e-mail: fubo.fu@gmail.com

P.P. Pande et al. (eds.), *Design Technologies for Green and Sustainable Computing Systems*,
DOI 10.1007/978-1-4614-4975-1_2, © Springer Science+Business Media New York 2013

Fig. 2.1 A simplified
diagram of network-on-chip

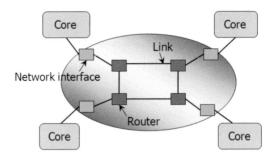

with technology scaling, the probability that a high-energy particle strike will flip
the logic value in a storage element increases [17]. The smaller pitch size produces
more crosstalk [18]. Moreover, the error rate of transient errors in logic gates is
expected to increase because of higher frequencies and lower supply voltages [19].
Therefore, research on the power efficiency and high reliability for the sustainable
computing systems is imperative.

The outline for the following section as follow: in Sect. 2.2, we overview the
common techniques used for reliable NoC design. In Sect. 2.3, several recent NoC
link design methods are presented. In Sect. 2.4, techniques for reliable NoC router
design are introduced. Summaries are provided in Sect. 2.5.

2.2 Overview for Reliable NoC Design

2.2.1 General Error Control Schemes

Three typical error control schemes are used in on-chip communication: error
detection combined with automatic repeat request (ARQ), hybrid ARQ (HARQ)
and forward error correction (FEC). The generic diagram for transmitter and
receiver is shown in Fig. 2.2. ARQ and HARQ use an acknowledge (ACK) or not
acknowledge (NACK) signal to request transmitter resending message; FEC does
not need ACK/NACK but try to correct error in receiver, although correction may be
wrong. In error detection plus automatic repeat request (ARQ) scheme, the decoder
in the receiver performs error detection. If an error is detected, retransmission is
requested. This scheme is proved as the most energy efficient method for reliable on-
chip communication, if the error rate is rarely small [20]. Hybrid automatic repeat
request (HARQ) first attempts to correct the detected error; if the error exceeds
the codec's error correction capability, retransmission is requested. This method
achieves more throughput than ARQ does, at the cost of more area and redundant
bits [21]. Extended Hamming code can detect and correct errors; thus, this code
can be employed to HARQ error control scheme. According to the retransmission
information, HARQ is divided into type-I HARQ and type-II HARQ categories

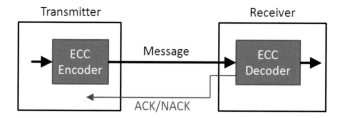

Fig. 2.2   Generic diagram for error control scheme

[21, 22]. The former one transmits both the error detection and correction check bits. In contrast, the latter one transmits parity checks for error detection. The check bits for error correction are transmitted only when necessary. As a result, type-II HARQ achieves better power consumption than type-I HARQ [23]. Forward error correction (FEC) is typically designed for worst-case noise condition. Different with ARQ and HARQ, no retransmission is needed in FEC [23]. The decoder always attempts to correct the detected errors. If the error is beyond the codec's capability, error correction is still performed. As a result, decoding failure occurs. Block FEC codes achieves better throughput than ARQ and HARQ; however, this scheme designed for worst-case condition wastes energy if the noise condition is favorable. Alternatively, encoding/decoding current input and previous input, convolutional code increases coding strength but yields significant codec latency [24]; thus, FEC with convolutional code is not suitable for on-chip interconnect network.

2.2.2   Error Control Coding

Error control coding (ECC) approaches have been widely applied ARQ, HARQ and FEC schemes mentioned above. In ECCs, parity check bits are calculated based on the input data. The input data and parity check bits are transmitted across interconnects. In the receiver, an ECC decoder is used to detect or correct the errors induced during the transmission. In early research work, simple ECCs, such as single parity check (SPC) codes, Hamming codes, and duplicate-add-parity (DAP) codes are widely used to detect or correct single errors. As the probability of multiple errors increases in nanoscale technologies, more complex error control codes, such as Bose-Chaudhuri-Hocquenghem (BCH) codes, Reed-Solomon (RS) codes and product codes are applied to improve the reliability of on-chip interconnects.

The single parity check (SPC) code is one of the simplest codes. In SPC codes, an additional parity bit is added to a k-bit data block such that the resulting $(k + 1)$-bit codeword has an even number (for even parity) or an odd number (for odd parity) of 1s. SPC codes have a minimum Hamming distance $d_{min} = 2$ and can only be used for error detection. SPC codes can detect all odd numbers of errors in a codeword.

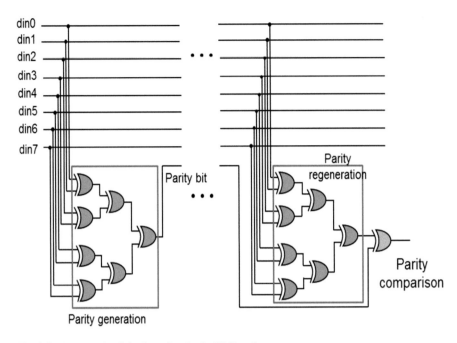

Fig. 2.3 An example of single parity check (SPC) codes

The hardware circuit used to generate the parity check bit is composed of a number of exclusive OR (XOR) gates as shown in Fig. 2.3. In the SPC decoder, another parity generation circuit, identical to that employed in the encoder, is employed to recalculate the parity check bit based on the received data. The recalculated parity check bit is compared to the received parity check bit. If the recalculated parity check bit is different from the received parity check bit, errors are detected. The bit comparison can be implemented using an XOR gate as shown in Fig. 2.3.

In duplicate-add-parity (DAP) codes [25, 26], a k-bit input is duplicated and an extra parity check bit, calculated from original data, is added. For k-bit input data, the codeword width of DAP codes is $2k + 1$. DAP codes have a minimum Hamming distance $d_{min} = 3$, because any two distinct codewords differ in at least three bit positions. It can correct single errors. The encoding and decoding process of DAP codes are shown in Fig. 2.4. In the DAP code implementation, each duplicated data bit is placed adjacent to each original data bit. Thus, DAP codes can also reduce the impact of crosstalk coupling.

Hamming codes are a type of linear block codes with minimum Hamming distance $d_{min} = 3$. Hamming codes can be used to either correct single errors or detect double errors. Figure 2.5 shows an example of the Hamming(7, 4) decoding circuits. The syndrome is calculated from the received Hamming codeword. The syndrome calculation circuit can be implemented as XOR trees. The calculated syndrome is used to decide the error vector through the syndrome decoder circuit.

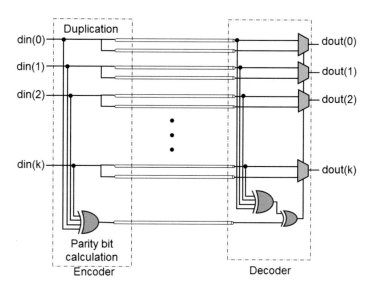

Fig. 2.4 Encoding and decoding process of DAP codes

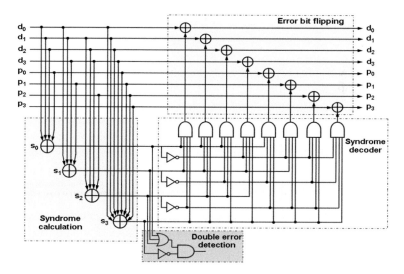

Fig. 2.5 Hamming (7, 4) decoder

The syndrome decoder circuit can be realized using AND trees. The syndrome and its inverse are the inputs of the AND trees. A Hamming code can be extended by adding one overall parity check bit. Extended Hamming codes have a minimum Hamming distance $d_{min} = 4$ and belong to single-error-correcting and double-error-detecting (SEC-DED) codes, which can correct single errors and detect double errors at the same time. Figure 2.6 shows an implementation example of the

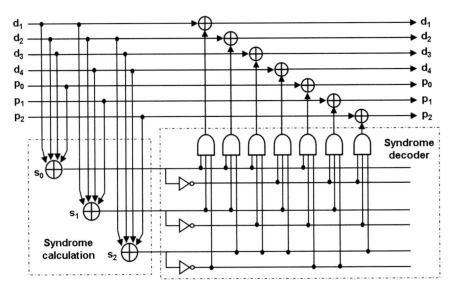

Fig. 2.6 Extended Hamming (8, 4) decoder

extended Hamming EH(8, 4) decoder. One of the syndrome bits is an even parity of
the entire codeword. If this bit is a zero and other syndrome bits are non-zero, this
implies that there were two errors—the even parity check bit indicates that there are
zero (or an even number of) errors, while the other non-zero syndrome bits indicate
that there is at least one error.

Hsiao codes are a special case of extended Hamming code with SEC-DEC capability. In Hsiao codes, the parity check matrix $H(n-k) \times n$ satisfies the following
four constraints—(a) Every column is different. (b) No all zero column exists. (c)
There are an odd number of 1's in each column. (d) Each row in parity check matrix
contains the same number of 1's. This parity check matrix has an odd number of
1's in each column and the number of 1's in each row is equal. The double error is
detected, when the syndrome is non-zero and the number of 1's in the syndrome is
not odd. The hardware requirement in the encoder and decoder of Hsiao codes is less
than that of extended Hamming codes, because the number of 1's in parity check
matrix of Hsiao codes is less than that in an extended Hamming code. Further, the
same number of 1's in each row of the parity check matrix reduces the calculation
delay of the parity check bits.

As the occurrence of multiple error bits is expected to increase with the
technology further scales down, error control for multiple errors has gained more
attentions than before. Interleaving is an efficient approach to achieve protection
against spatial burst errors. Multiple SEC codes combined with interleaving can
correct spatial burst errors [27, 28]. In this method, the input data is separated into
small groups. Each group is encoded separately using simple linear block codes
(e.g., SEC codes). Another method to detect burst errors is cyclic redundancy code
(CRC), which is a class of cyclic code [24]. The encoding process of cyclic codes

can be realized serially by using a simple linear feedback shift register (LFSR). The use of an LFSR circuit requires little hardware but introduces a large latency when a large amount of data is processed. Cyclic codes can also be encoded by multiplying input data with a generator matrix. BCH codes are an important class of linear block codes for multiple error correction [24]. RS codes are a subclass of nonbinary BCH codes [24, 29] that are good at correcting multiple symbol errors.

2.2.3 Fault Tolerant Routing

Other than error control coding, NoCs are capable of employing fault tolerant routing algorithms to improve error resilience. The fault tolerant capability is achieved by using either redundant packets or redundant routes. In redundant-packet-based fault tolerant routing algorithms, multiple copies of packets are transmitted over network, so that at least one correct packet can reach the destination. The disadvantages of this routing category: (1) add more network congestion; (2) increase power consumption; (3) fault tolerance capability decreases if the number of copies decreases; (4) boost the router design complexity. Different efforts have been made to improve the efficiency of redundant-packet-based routing. Flooding routing algorithm requires the source router sending a copy of the packet to each possible direction and intermediate routers forwarding the received packet to all possible directions as well [30]. Various flooding variants have been proposed. In probabilistic flooding, source router sends copies of the packet to all of its neighbors and the middle routers forward the received packets to their neighbors with a pre-defined probability (a.k.a gossip rate), which reduces the number of redundant packets [31]. In directed flooding, the probability of forwarding packet to the particular neighbor is multiplied with a factor depending on the distance between current node and the destination [32]. Different with previous flooding algorithms, Pirretti et al. proposed a redundant random walk algorithm, in which the intermediate node assigns different packet forwarding probabilities to the output ports (but the sum is equal to 1). As a result, this approach only forwards one copy of the received packet, reducing the overhead [33]. Considering the tradeoff of redundancy and performance, Patooghy et al. only transmit an additional copy of the packet through low-traffic-load paths for replacing the erroneous packet [33]. Redundant-packet-based routing is feasible for both transient and permanent errors, no matter whether the error presents in links, buffers or logic gates in the router.

In redundant-route-based fault tolerant routing algorithms, single copy of the packet is transmitted via one of the possible path. This category routing algorithm takes advantage of either global/semi-global information or distributed control to use NoC inherent redundant routes in topology for handling faults. Representative redundant-route routing using global/semi-global information are distance vector routing, link state routing, DyNoC and reconfigurable routing. In distance vector routing, the number of hops between current router and each destination are periodically updated in the routing table. As a result, the faulty link and router can be notified in each router within one period. Link state routing uses handshaking protocol to sense the state of neighbor links and router, so that the faulty links and routers can be considered in computing the shortest path [33]. Unlike distance vector routing and link state routing, dynamic NoC routing does not broadcast the broken links or permanently unusable routers to the entire network; instead, only neighboring routers receive the notification, so that the obstacle can be bypassed [34]. In reconfigurable routing [35], eight routers around the broken router are informed, and those routers use other routing path to avoid that router and prevent the presence of deadlock.

Global control routing algorithms are aiming to obtain the optimal path, but resulting in large area overhead, power consumption and design complexity. In contrast, distributed control routing algorithms have fewer overheads than global ones; they only gather information from their directly connected routers, thus not always optimal. A large portion of distributed control routing algorithms [36–38] is used to avoid network congestion or deadlock. Recently, those algorithms have been employed to tolerate permanent faults [35–41]. In contrast, Zhou and Lau [42], Boppana and Chalasani [43], and Chen and Chiu [44] took advantage of virtual channels to extend the region of re-routing for the flits encountering fault interference. The fault-tolerant adaptive routing algorithm proposed by Park et al. in [45] requires additional logic circuit and multiplexers for buffers and virtual channel switching, respectively. The routing algorithms analyzed by Duato need at least four virtual channels per physical channel, which is not desirable for area-constraint NoCs [46]. In [41], Schonwald et al. proposed a force-directed wormhole routing algorithm to uniformly distribute the traffic across the entire network, in the process of handling fault links and switches. Link permanent errors can also be addressed by adding spare wires [47, 48].

2.3 Reliable NoC Link Design

2.3.1 Energy Efficiency ECC

A powerful ECC usually requires more redundant bits and more complex encoding and decoding processes, which increases the codec overhead. To meet the tight speed, area, and energy constraints imposed by on-chip interconnect links, ECCs used for on-chip interconnects need to balance reliability and performance.

2.3.1.1 Hamming Product Codes

Product codes were first presented in 1954 [49]. The concept of product codes is very simple. Long and powerful block codes can be constructed by serially concatenating two or more simple component codes [49–51]. Figure 2.7 shows

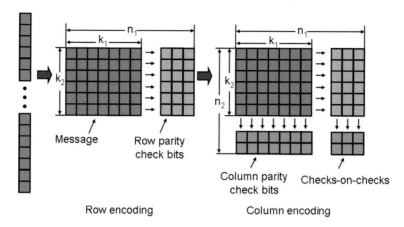

Fig. 2.7 Encoding process of product codes

the construction process of two dimensional product codes. Assume that two component codes $C_1(n_1, k_1, d_1)$ and $C_2(n_2, k_2, d_2)$ are used, where n_1, k_1 and d_1 are codeword width, input data width, and minimum Hamming distance for the code C_1, respectively; n_2, k_2 and d_2 are codeword width, input data width, and minimum Hamming distance for the code C_2, respectively. The product code $C_p(n_1 \times n_2, k_1 \times k_2, d_1 \times d_2)$ is from C_1 and C_2 as follows:

1. Arrange input data in a matrix of k_2 rows and k_1 columns.
2. Encode the k_2 rows using component code C_1. The result will be an array of k_2 rows and n_1 columns.
3. Encode the n_1 columns using component code C_2.

Product codes have a larger Hamming distance compared to that of the component codes. If the component codes C_1 and C_2 have minimum Hamming distance d_1 and d_2 respectively, then the minimum Hamming distance of the product code C_p is the product $d_1 \times d_2$, which greatly increases the error correction capability. Product codes can be constructed by a serial concatenation of simple component codes and a row-column block interleaver, in which the input sequence is written into the matrix row-wise and read out column-wise. Product codes can efficiently correct both random and burst errors. For example, if the received product codeword has errors located in a number of rows not exceeding $(d_2 - 1)/2$ and no errors in other rows, all the errors can be corrected during column decoding.

The Hamming product codes can be decoded using a two-step row-column (or column-row) decoding algorithm [24]. Unfortunately, this decoding method fails to correct certain error patterns (e.g. rectangular four-bit errors). A three-stage pipelined Hamminproduct code decoding method is proposed in [52]. Compared to the two-step row-column decoding method, the three-stage pipelined decoding method uses a row status vector and a column status vector to record the behaviors of the row and column decoders. Instead of passing only the coded data between

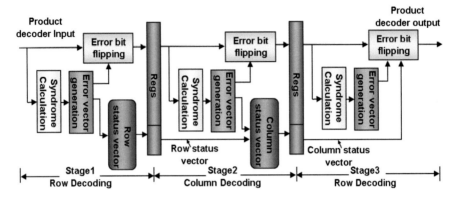

Fig. 2.8 Block diagram of three-stage pipelined decoding algorithm

row and column decoder, these row and column status vectors are passed between stages to help make decoding decisions [52]. The simplified row and column status vector implementation can be described as follows: The ith $(1 \leq i \leq n_2)$ position in the row status vector is set to "1" when there are detectable errors (regardless of whether the errors can be corrected or not) in the i_{th} row; otherwise that position is set to "0". For the column status vectors, there are two separate conditions that can cause the jth $(1 \leq j \leq n_1)$ position in column status vector to be set to "1" (a) when an error is detectable but not correctable, or (b) when an error is correctable, but the row where the error occurs has a status value "0". Otherwise, that position is "0".

Figure 2.8 describes the three-stage pipelined Hamming product code decoding process. After initializing all status vectors to zeros, the steps are described as follows: Step 1: Row decoding of the received encoded matrix. If the errors in a row are correctable, the error bit indicated by the syndrome is flipped. The row status vector is set to "0" if the syndrome is zero and "1" if the syndrome is nonzero. Step 2: Column decoding of the updated matrix. The error correction process is similar to Step 1. The column status vector is calculated using both the column error vector and the row status vector from Step 1. Step 3: Row decoding the matrix after changes from Step 2. The syndrome for each row is recalculated. If any remaining errors in each row are correctable, the row syndrome will be used to do the correction. If the errors in each row are still detectable but uncorrectable, the column status vector from Step 2 is used to indicate which columns need to be corrected.

To balance complexity and error correction capability, an error control method combining extended Hamming product codes with type-II HARQ is introduced by [52]. The encoding process of the combination of extended Hamming product codes with type-II HARQ is simple. In the decoding process of extended Hamming product codes with type-II HARQ, the received data is first decoded row by row using multiple extended Hamming decoders. Extended Hamming codes can correct single errors and detect double errors in each row. If all errors are correctable (no more than one error in each row), the receiver indicates a successful transmission by

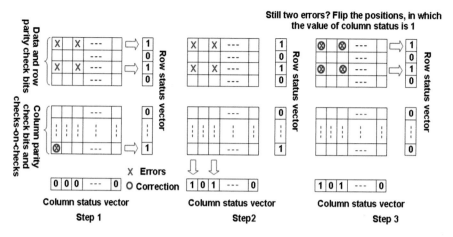

Fig. 2.9 An example of decoding process for extended Hamming codes with type-II HARQ

sending back an ACK signal to transmitter. If the receiver detects two errors in any row, it saves the row decoded data and row parity check bits in the decoding buffer and requests a transmission of column parity check bits and checks-on-checks by sending back a NACK signal. When the extra parity check bits are received, they are used with the saved data and row parity check bits to complete the column decoding process and the second row decoding process in the three-stage pipelined decoding method. Figure 2.9 shows an example of the decoding process when extended Hamming product codes with type-II HARQ are applied. A rectangular four-error pattern occurs in the transmission of the original data and row parity check bits. The extended Hamming decoder detects these errors during the first row decoding process and a transmission of column parity check bits and checks-on-checks is requested. A single error occurs during retransmission of column parity check bits and checks-on-checks. It can be directly corrected, before these extra parity check bits are combined with the saved data and row parity check bits to complete the three-stage pipelined decoding process. In Step 2, because double errors are detectable but uncorrectable, no correction is performed and "1"s are recorded in the corresponding column states. In the second row decoding process (Step 3), the extended Hamming decoder still detects two errors in a row, so the column status vector is used to indicate which positions need to be flipped.

2.3.1.2   Experimental Results

Reliability

The residual flit error rate $P_{residual}$ is used to measure the system reliability [52]. Figure 2.10 shows the residual flit error rate of different error control schemes as a function of noise voltage deviation. Dependent error model introduced in [52] is

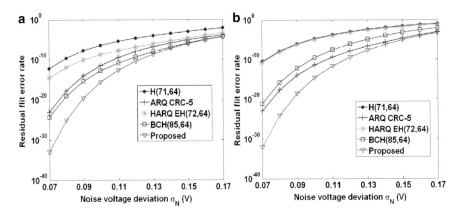

Fig. 2.10 Residual flit error rate for different error control schemes as a function of noise voltage deviation at (**a**) $P_n = 10^{-2}$ and (**b**) $P_n = 1$

used in the simulation with two coupling probability values, $P_n = 10^{-2}$ and $P_n = 1$. A link swing voltage of 1.0 V is used. The simulation results show that Hamming product codes with type-II HARQ achieves a significant reduction in residual flit error rate when multiple random and burst errors are considered. ARQ CRC-5 has a good burst error detection capability but it is inefficient to detect multiple random errors. HARQ EH(72, 64) scheme can correct single errors and detect double errors but as the burst error probability increases, the performance of decreases. Compared to the BCH(85, 64) code, extended Hamming product codes with type-II HARQ can effectively correct multiple random and burst errors, while BCH code is only good at correcting multiple random errors. The combination of extended Hamming product codes with type-II HARQ can correct at least two permanent errors, while ARQ CRC-5 will not work in this persistent noise environment.

Energy Consumption

The average energy per flit is used as the metric to measure energy consumption. The average energy consumption includes the encoder energy E_{e1}, the link energy E_{l1}, and the decoder energy E_{d1} in the first transmission, as well as the encoder energy E_{e2}, the link energy E_{l2}, and the decoder energy E_{d2} in the retransmission,

$$E_{avg} = (E_{e1} + E_{l1} + E_{d1}) + P_{d\_uc}(E_{e2} + E_{l2} + E_{d2}) \qquad (2.1)$$

where P_{d_uc} is the probability that the errors are detectable but uncorrectable. The link energy using low swing voltage can be estimated as,

$$E_l \approx S_f \cdot W_L \cdot C_L \cdot V_{DD} \cdot V_{swing} + E_{level} \qquad (2.2)$$

where C_L is the interconnect capacitance. W_L is the number of wires in the link, which depends on the error control scheme. In the combination of extended

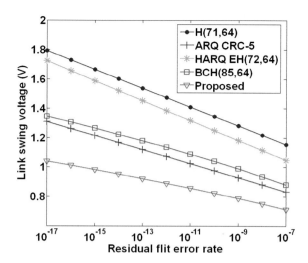

Fig. 2.11 Required link swing voltages of different error control schemes for given reliability requirements

Hamming product codes with type-II HARQ, W_L is greatly affected by the selection of k_2. S_f is the wire switching probability. V_{DD} is the supply voltage. The link swing voltage V_{swing} is decided by the reliability requirement. E_{level} is the energy consumption of the level translation circuit when low swing voltage is applied.

Figure 2.11 compares the link swing voltage of different error control schemes for the same residual flit error rate requirement. The σ_N is assumed to be 0.1 V. The coupling probability P_n is 10^{-1}. The results show that the more effective the error correction capability of an error control scheme, the lower the swing voltage needed for the interconnect links. To achieve the same residual flit error rate, the combination of extended Hamming product codes with type-II HARQ achieves the smallest link swing voltage. The link swing voltage of the combination of extended Hamming codes with type-II HARQ is about 60% and 80% compared to that of the H(71, 64) and ARQ CRC-5, respectively. The lower link swing voltage allows this method to consume less link energy.

Figure 2.12 compares the link energy consumption of different error control schemes given the same residual flit error rate requirement. In the simulation, the requirement of residual flit error rate is assumed to be $P_{residual} \leq 10^{-20}$. The simulation was performed for a noise environment of $\sigma_N = 0.07$ V. Different technology nodes are considered in the simulation using Predictive Technology Model (PTM) CMOS 65 and 45 nm technology. The effect of different link lengths on energy consumption is also evaluated. In NoCs, the link length is the distance between two routers, which is decided by the dimension of the tile block. In mesh or torus topologies, the links between two routers are generally a few millimeters long wires. In the experiments, link lengths from 1 to 3 mm are examined. The link energy is measured in Cadence Spectre. The input data is generated using an H.264 video encoder with the average switching factor about 0.5. Figure 2.12 shows that the combination of extended Hamming product codes with type II HARQ has the smallest link energy of the compared schemes, because the lowest link

Fig. 2.12 Link energy
consumption of different
error control schemes for
different link lengths (a)
45 nm technology (b) 65 nm
technology

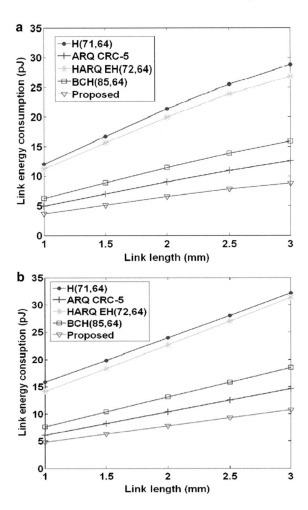

swing voltage counterbalances the large number of wires in the link. As link length
increases, Hamming product codes with type II HARQ can benefit more from the
lowest link swing voltage. The link energy consumption of Hamming product codes
with type-II HARQ is about 80% and 35% compared to the link energy consumption
of ARQ CRC-5 and H(71, 64), respectively.

Figure 2.13 compares the average energy consumption per flit for different
error control schemes at the same reliability requirement (10^{-20}). The average
energy includes encoder, decoder and link energy consumption. Two noise voltage
deviations, $\sigma_N = 0.07$ V and $\sigma_N = 0.1$ V, are considered. Raw bit error probability
ε is about 10^{-12} and 10^{-6} for these two cases. The results show that ARQ
CRC-5 achieves the least average energy consumption at low noise environment
($\sigma_N = 0.07$ V) for small link lengths, because of the smaller codec energy and link
energy consumption. As the noise voltage deviation increases, however, higher link

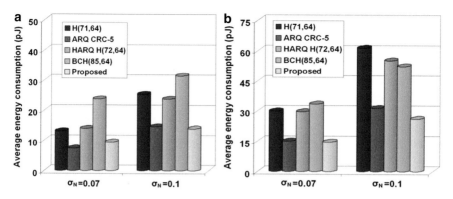

Fig. 2.13 Energy comparison of different error control schemes at residual flit error rate 10^{-20} (**a**) Link length 1 mm (**b**) Link length 3 mm

swing voltages are needed to achieve the same reliability. In high noise conditions, the average energy consumption of ARQ CRC-5 increases more than the average energy consumption of the combination of extended Hamming product codes with type-II HARQ, because ARQ CRC-5 has larger link energy consumption. The combination of extended Hamming product codes yields the least average energy consumption at the higher noise environment ($\sigma_N = 0.1$ V). The BCH(85, 64) scheme has the larger average energy consumption for small link lengths because it has the largest codec energy consumption. Hamming product codes with type-II HARQ achieve the least energy consumption at large link lengths or high noise environments. When the link length is 3 mm, the energy consumption of the this approach is about 15% and 50% less than that of ARQ CRC-5 and H(71, 64), respectively, in high noise environment. In addition to the energy consumption improvement compared to ARQ in high noise environments, the combination of extended Hamming product codes with type-II HARQ can correct at least two permanent errors, while ARQ will not work in a persistent noise environment. Thus, the approach combining forward error correction with limited retransmission can achieve a better performance for balanced energy, performance, and error resilience.

2.3.2 Combining Error Control Codes with Crosstalk Reduction

2.3.2.1 Crosstalk Avoidance Codes

DAP codes can reduce capacitive coupling [25, 26]. An intelligent spacing method can be used to further optimize DAP code [53]. In intelligent spacing method, the spacing between two wires carrying the identical data can be smaller than the spacing between two wires carrying different data. BSC [54] does not have any adjacent bits simultaneously switch in opposite direction (i.e., no $01 \to 10$ or $10 \to 01$ transition at two adjacent bit positions). The DAP code concept can be extended to construct Crosstalk Avoidance and Multiple Error Correction Code (CAMEC) codes [55], in which the input data is first encoded using Hamming codes, and the outputs of the Hamming encoder are duplicated and an overall parity check bit, calculated from the output of the Hamming encoder, is added to the whole codeword. The CADEC decoding algorithm can only guarantee to correct double errors. An updated joint crosstalk avoidance and triple error correction (JTEC) decoding algorithm is introduced in [56]. The JTEC code can guarantee to correct three bit errors. In JTEC codes, the Hamming code along with the overall parity bit comprise of an extended Hamming code, which can correct single error and detect double error at the same time. A unified coding framework by combining error control coding with crosstalk avoidance codes is proposed in [25, 57]. In this method, the input data is first encoded using nonlinear crosstalk avoidance codes (CACs). The outputs of CACs are encoded using an error control code. The parity bits generated by the error control codes are protected against crosstalk coupling using techniques such as shielding and duplication. There are three common used CACs—forbidden overlap condition (FOC) codes [58], forbidden transition condition (FTC) codes [59], and forbidden pattern condition (FPC) codes [60]. Each of these CACs has different crosstalk reduction capabilities.

2.3.2.2   Error Control Codes with Skewed Transitions

Skewed transition method [61, 62] is used to reduce crosstalk coupling by delaying adjacent transitions with some finite time ΔT. In this section, we will introduce another method combining error control coding with skewed transitions to simultaneously address error correction and capacitance coupling induced delay uncertainty. In skewed transitions, the simultaneous opposite switching on neighboring bus lines are avoided by the induced relative delay ΔT. The worst-case effective capacitance C_{eff} of skewed transition (C_{eff} of a middle wire when a $010 \to 101$ or $101 \to 010$ transition occurs on three adjacent wires) can be expressed by Eq. 2.3 below [61],

$$C_{eff} = C_{gt} + \left(4 - 2\frac{|V_N(\Delta T) - V_N(0)|}{V_{DD}}\right)C_{ct}$$
$$= C_{gt} + (4 - 2v(\Delta T))C_{ct} \qquad (2.3)$$

where C_{gt} is the total capacitance between the wire and ground; C_{ct} is the total coupling capacitance between any two adjacent wires; $V_N(\Delta T)$ and $V_N(0)$ are the voltages of neighboring wires at time ΔT and 0, respectively; $v(\Delta T)$ is the ratio of the neighboring wire's voltage difference at time ΔT and 0 to V_{DD} ($0 \leq v(\Delta T) \leq 1$). When $\Delta T = 0$, v is 0. As ΔT increases, v approaches 1.

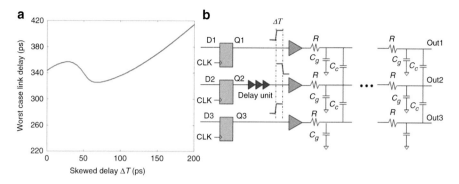

Fig. 2.14 Conventional skewed transitions (**a**) Skewed transition by inserting delay elements (**b**) The relation between the worst-case link delay and skewed delay ΔT

In skewed transition methods, delay elements are inserted at the beginning of alternate bus lines to generate the relative delay ΔT, as shown in Fig. 2.14a. For a bus line with $k-1$ repeaters, the worst-case link delay T_d in skewed transitions can be described by Eq. 2.4 below [61],

$$T_d = \left(0.7R_r + 0.4\frac{R_t}{k}\right)(C_{gt} + 4C_{ct}) + 0.7(kR_r + R_t)C_r + \Delta T$$

$$- \left(1.4R_r + 0.8\frac{R_t}{k}\right)C_{ct}v(\Delta T) \tag{2.4}$$

where R_r and C_r are the on-resistance and output capacitance of the repeater. R_t is the total resistance of the wire. The first two terms in Eq. 2.4 are the worst-case delay of the standard bus. From Eq. 2.4, the delay reduction achieved by the skewed transition method depends on the difference between the last two terms. Thus, a large ΔT increases the overall link delay T_d, as shown in Fig. 2.14b.

In [63], a method combining ECCs and skewed transitions is proposed to improve the reliability of on-chip interconnects. In this method, ECCs are used to correct logic errors while skewed transitions are applied to reduce capacitive crosstalk induced delay uncertainties. By hiding the delay insertion overhead of the skewed transition method, this method achieves a larger reduction in the worst-case link delay compared to conventional skewed transition method. Figure 2.15 shows the method combining ECCs with skewed transitions. In an error control encoder, the parity bits are generated from the original input data after a finite delay. Instead of sending the input data and parity bits to the link at the same time, partial input data can be sent before the parity bits are available. Two clocks (*CLK1* and *CLK2*, with *CLK1* arriving ahead of *CLK2*) are used alternately to offset the transitions in each pair of adjacent interconnect lines. The input data and parity bits are mapped to registers triggered by these two clocks, as shown in Fig. 2.15.

Figure 2.16 illustrates the transmission procedure of the method combining ECCs with skewed transition. Assume that the clock cycle of *CLK1* and *CLK2* is

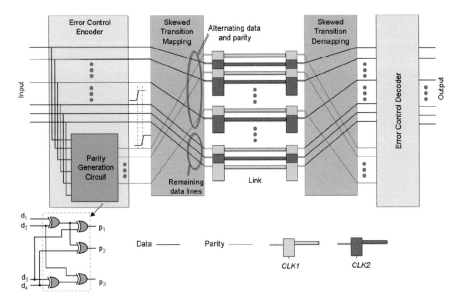

Fig. 2.15 Block diagram of proposed method exploiting parity computation latency to reduce crosstalk coupling

T_{cycle} and k-bit input data are available at the rising edge of *CLK1*. The calculation of r-bit parity data is completed after a delay of D_{parity}. The wires $l(i)$ ($1 \leq i \leq k+r$) in the link with odd index i are triggered by *CLK1* and $l(i)$ with even index i are triggered by *CLK2*. In the proposed method, input data can be sent at the next rising edge of *CLK1* or at the rising edge of *CLK2*, which arrives after a delay ΔT_1, as shown in Fig. 2.16. Because the data bits are available before the parity bits are calculated, thus the data can be sent earlier than the parity bits without affecting the overall system performance. Parity-check bits are calculated using the input data after the delay D_{parity}; thus, they can only be transmitted at the next rising edge of *CLK1*. The relationship between T_{cycle} and the timing offsets ΔT_1 and ΔT_2 is described in Fig. 2.16 and should meet the following constraint Eq. 2.5. For implementation simplicity, CLK1 and CLK2 can be the rising and falling edge of the same clock.

$$T_{cycle} = \Delta T_1 + \Delta T_2 \geq D_{parity} \qquad (2.5)$$

2.3.2.3 Experimental Results

Performance Evaluation

Unlike conventional skewed transition methods, the combination of error control codes with skewed transitions hides the overhead induced by delay elements

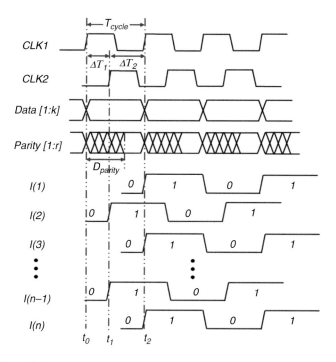

Fig. 2.16 Transmission procedure of the method combining error control coding with crosstalk reduction

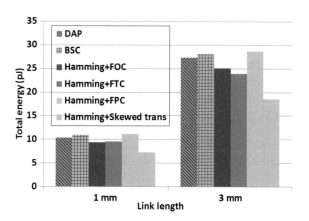

Fig. 2.17 Total energy versus link length for different schemes simultaneously addressing error correction with crosstalk reduction

in the ECC encoding stage. Figure 2.17 compares the worst-case link delay of the combination of ECCs with skewed transitions with the conventional skewed transition method. A Hamming H(71, 64) code is used to correct single logic errors. The Hamming encoder is implemented as XOR trees. The depth of the XOR trees determines the worst-case delay of the Hamming encoder. The H(71, 64) encoder is synthesized using a TSMC 65 nm technology with the worst-case delay $D_{parity} = 400$ ps. A 65 nm link model [64] with lengths from 1 to 5 mm is used in the simulations. The link delay is normalized to the delay of a standard bus with minimum link width and spacing. Figure 2.17 compares the total energy consumption E_{total} of each method at link lengths of 1 and 3 mm. E_{tota} includes encoder, link, and decoder energy. The results show that combining H(71, 64) with FPC consumes more total energy than other schemes, because of the larger codec and link energy consumption. The combination of H(71, 64) with skewed transitions achieves the least total energy consumption because of the relatively small codec overhead and the least required number of wires. Compared to combining H(71, 64) with FPC, it can achieve 32% improvement in energy consumption at link length 3 mm.

2.4 Reliable NoC Router Design

2.4.1 Router Architecture

Typically, error control coding techniques are only effective in linear systems. Unfortunately, the router control path is not a linear system, requiring alternative approaches to manage general routing errors. One common approach to handle permanent errors in router is fault tolerant routing. These methods involve isolating the entire router or a few ports of a router [65–67] if permanent errors are detected. This isolation creates irregularities in the NoC topology, potentially degrading NoC performance. To manage permanent errors in router, we can either add spare components to replace the defective elements or increase the burden on the remaining usable elements in the system. Both of those solutions achieve fault tolerance at the cost of increasing area overhead, degrading performance, or consuming more energy. In addition, the permanent error management methods generally are based on the assumption that the faulty components in the router have been recognized in the test phase. However, transient errors happen during the runtime and cannot be predicted. Consequently, new energy-efficient and reliable methods are imperative to control transient errors in routers.

Triple-modular redundancy (TMR) duplicates the unit under protection and determines the most likely output by using a majority voting. Because of its simplicity, TMR has been applied to the router control paths [53, 68]. Theoretically, TMR functions correctly when up to one-third of the received copies are wrong. Error happened to the majority voter further reduces the effectiveness of the TMR approach, particularly when the size of the unit under protection is small [69]. Consequently, TMR is not an ideal solution for the control paths in NoC routers.

In this section, we focus on transient error management for the NoCs using popular mesh topology and five-port routers. To obtain high clock frequency, we assume each hop has four pipeline stages: three for the router and one for links between two neighboring routers, as shown in Fig. 2.18. Note that each dash-dotted

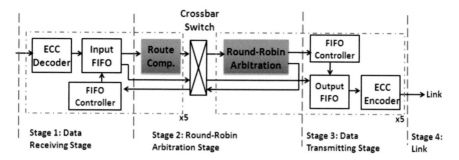

Fig. 2.18 Generic pipelined router architecture

box is duplicated by five times in each router, because five-port router is interested here. In the first stage of the router, ECC decoder is used to manage interconnect link errors. Error-free packets are stored in the input FIFO. In the second stage, the route computation (RC) block extracts the type of the incoming flit, determines the desired output port from the given destination address, and requests access to the appropriate output port. The round-robin arbitration (RRA) unit is composed of round-robin computation unit and registers for priority vectors and port reservation information. Each RRA unit grants the input port with the highest-priority to access the output port that follows that RRA unit. In the third stage, flits popped out from the output FIFO are encoded before transmission. In this section, we concentrate on the error resilience of *Stage 2*, the router control path, to complementary to error control methods for link and router data path.

The inherent information redundancy is obtained through the occurrence of forbidden signal patterns or inconsistent request-response pairs in the system. Note that failure detection in this work is applied to dimensional XY routing. In this work, we use a practical packet format [70]. Each packet has one header flit, one tail flit and several payload flits. The first two bits of a header flit are '10'. The remaining bits in the header flit contain information such as source identifier, destination coordinator and routing protocol. The first two bits of a tail flit are '01'. A flit with high logic on both the first and second bits is a payload flit.

2.4.2   Reliable Router Architecture

Error correction typically results in more overhead than error detection. We proposed a method that provides error correction to the unit only if errors in that unit will directly result in packet loss or misrouting [71]. As shown in Fig. 2.19, the RC unit is protected by error detection unit (1). If RC computation errors are detected, a warning signal is activated to stop the input FIFO popping out the next flit and to request RC re-computation. This re-computation process results in additional one-cycle latency, but it is faster than multi-cycle rerouting The warning signal is also

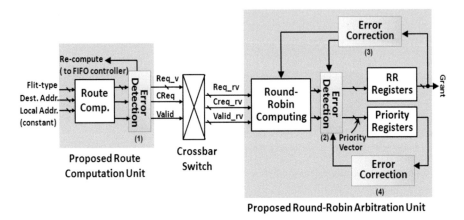

Fig. 2.19 Block diagram of proposed round robin arbiter. We propose the addition of the four shaded units to the conventional arbiter design, enabling error control in the arbitration. *RC* route computation, *RR* round-robin

used to clear the request of using one of the output ports, with the logic—e.g. *Req_v* & (*Re-compute*). The *Req_v* vector after error detection unit (1) informs output ports whether a new packet header arrives. The *CReq* signal indicates to release the reserved input-output port connection in the next cycle. The *ValidFlit* vector indicates if the current flit is a valid flit (i.e., header, payload, or tail flit). The RRA unit in Fig. 2.18 is composed of round-robin computation (RRC) unit and registers shown in Fig. 2.19. The round-robin (RR) registers store the grant vector used in the last cycle. Priority registers save the priority vector, which will instruct the RRC unit to select the next highest-priority input port. By exploiting inherent information redundancy, we use four separated error management components—error detection unit (2), error correction units (3) and (4)—to prevent spatial and temporal error propagation. In our implementation, the four error control units are merged with the RC and RRC units to reduce the critical path delay.

2.4.3 Route Computation

2.4.3.1 Failures in Route Computation (RC) Unit

The function of the RC unit is to determine which output port the current input port should connect to. This means no more than one output port is requested each time, and this exclusive feature is regarded as one of the inherent information redundancies. According to RC unit functionality, three request failures happen:

1. *Mute-request*: No output port request when a header flit arrives at the input port. This failure does not result in packet loss, but delays the release of the current input port, increasing latency.

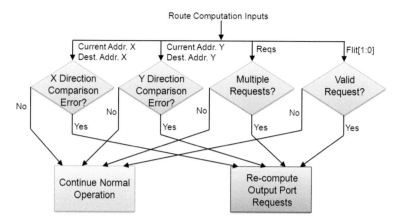

Fig. 2.20 Operation flowchart of proposed route computation unit

2. *Multiple-request*: One header flit requests multiple output ports at one cycle (we assume no flooding protocol is being used here). This type failure will result in multi-copy transmission, which consumes extra energy and potentially increases the network traffic congestion.

3. *Request-switch*: A non-header flit arrives but there is a request to build a new input-output port connection; or the request of the intended output port is muted while there is a simultaneous request for another output port. Although the RC unit only produces a single request, this erroneous request results in packet misrouting and even deadlock.

Detecting and correcting these three failures can effectively prevent error propagation to the next hop, saving energy on unnecessary network fabrics and logic gate switching.

2.4.3.2 Sigma and Branch Detection Method

A sigma and branch detection method is introduced detect mute-request, multiple-request and request-switch failures [71]. The flowchart for the proposed method is shown in Fig. 2.20. Any detected failure turns the warning signal to high, demanding re-calculation of the output port requests. For XY routing, the route computation (RC) unit compares the destination address, indicating by the coordinator (X_D, Y_D) in the 2D mesh network, with the current node ID represented by (X_C, Y_C).

Figure 2.21a shows the diagram of the request generation circuit in the RC unit. The number of active outputs (i.e., $Req\_E/S/W/N/L$) cannot be more than '1's at each computation. There exists two pairs of branch points—(A_0, A_1) and ($B0$, $B1$), which have forbidden the pattern '11'. We propose to warn the system by examining the number of active Req signals and checking the forbidden status of the branch points. If the number of requests is more than one, the branch points have the same

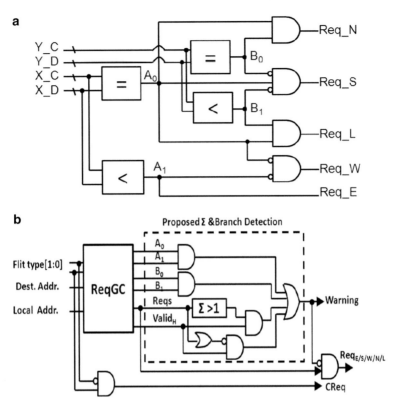

Fig. 2.21 Sigma & branch detection method. (**a**) Requests generation circuit (ReqGC), (**b**) Error detection circuit

high logic value or invalid request arrives, a warning signal is turned into high. The proposed architecture is shown in Fig. 2.21b. A sigma function ($\Sigma > 1$) is applied to detect the multiple-request events. The validation of the incoming flit is examined to detect the muted request. The complementary status of the branch points is checked to capture the failure behavior.

2.4.3.3 Evaluation

The influence of the number of failed gates on the route computation (RC) unit reliability is examined in Fig. 2.22. We randomly inject errors to the RC unit netlist. The output of each gate is possible to be flipped. Since the number of logic gates in the RC unit is less than 50, each data point is obtained by averaging 50,000 random simulations. As shown in Fig. 2.22, examining the number of requests (i.e., Σ detection) can reduce the system failure rate to 0.2. Combining Σ detection with branch point detection can further reduce the RC failure rate. Compared to the TMR method, our method achieves smaller system failure rate. This is valid even if only a single gate fails in the RC unit.

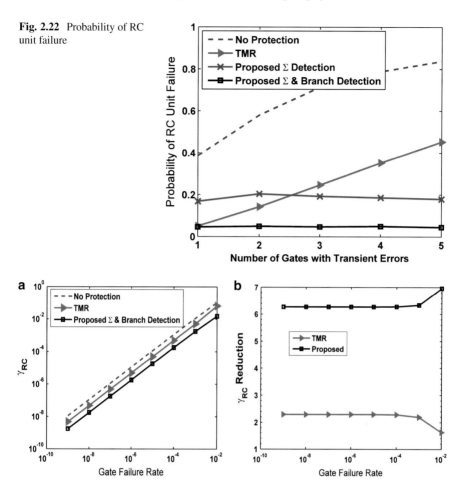

Fig. 2.22 Probability of RC unit failure

Fig. 2.23 (a) Impact of gate failure rate on RC unit failure rate, (b) RC unit failure rate reduction

Figure 2.23a shows the relationship of error management method and gate failure rate with the system failure rate γ_{RC}. TMR cannot reduce the system failure rate by several orders of magnitude as expected. This is because the duplicated RC unit has the same probability of experiencing logic gate failure. Moreover, TMR has large area overhead, thus it is more likely to have errors. Our method considers both the error detection and correction capability (i.e., α) and the area cost, providing a smaller system failure rate (γ_{RC}) than TMR. Figure 2.23b clarifies the RC unit failure rate reduction over the no-protection case. As can be seen, our method achieves three times system failure rate reduction than the TMR, and yet, the overhead of TMR is 2.6X over our method, as shown in Fig. 2.24. The limitation of our method is 10% more latency on the critical path, compared with TMR.

Fig. 2.24 Overhead
comparison

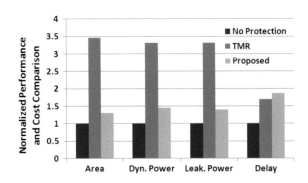

2.4.4 Arbitration Unit

2.4.4.1 Failures in Arbitration Unit

Each packet experiences three steps—create an input port-output port connection, maintain the port reservation and release the connection—to transfer a packet over the RRA unit. The RRA unit in each port grants a single connection between that port and one of the input ports. The connection remains until the request signal *CReq* is high. The RRA unit may experience four types of failures:

1. *Valid header flit with already active grant*: the round-robin computation unit changes the priority vector after each packet transmission. If one or more bits in the RR and priority registers are flipped by particle strikes, an input port may be given the grant before the header flit requests.
2. *Valid payload flit without active grant*: the header flit has reserved one output port; however, the reservation information is corrupted. Consequently, the payload flits cannot be continually transferred and network congestion occurs.
3. *Valid tail flit without active grant*: the tail flit cannot be forwarded to the output port and the input port-output port connection cannot be released. This also cause flit loss and network congestion.
4. *Invalid incoming flit with active grant*: the output port provides a grant to an input port, in which no valid flit exists. This grant cannot be cleared unless another failure happens in the RRA unit.

These inconsistent request-response pairs are used to provide error resilience in the RRA unit, preventing packet corruption and loss.

Round-robin arbitration (RRA) unit is used to reserve an output port for one input port when a header flit arrives. In the RC unit, transient errors can be managed with re-computation and the residual errors do not result in packet loss. However, the undetected errors in the RRA unit will lead to packet loss and network congestion increase. Consequently, the error management for RRA needs error correction capability.

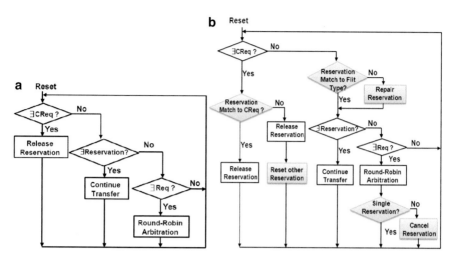

Fig. 2.25 Flowchart of round-robin arbitration. (**a**) Generic and (**b**) Proposed

2.4.4.2 Self-Correcting Method

Generic flowchart for the error-free arbitration unit is shown in Fig. 2.25a. The presence of *CReq* has the highest priority, because *CReq* releases the reserved resource to transfer new packets over the router. If one of the reservation registers is not zero, no new request can be granted until the current packet transmission completes. Round-robin arbitration is executed only when the output port is available. However, the arbitration unit composed of logic gates and storage elements may be affected by the voltage fluctuation and particle strikes, experiencing mute reservation, switching reservation and multiple reservations similar to destination computation unit. We propose a new RRA unit, which exploits the inherent information redundancy to perform error detection or/and correction. Assume *Req*, *CReq* and *flit type* information are error free because of the error management in the RC unit. We examine the errors in reservation registers (i.e., round-robin registers), by checking the consistency between the register contents and the *Req*, *CReq* and *flit type* signals. The proposed flowchart is shown in Fig. 2.25b. Three error management operations are added to the branches shown in Fig. 2.25a.

Conflict Between Output Port Reservation and CReq

We examine whether the content in the RR registers matches to the CReqs or not before updating the registers (i.e., releasing output port reservation). If no error, only the input port that reserves the current output port issues a valid CReq; any incident termination on the output port reservation is caused by errors. This constraint is the inherent information that can be used to correct the corrupted round-robin registers. The incorrect registers are reset when the reservation does not match to the CReq.

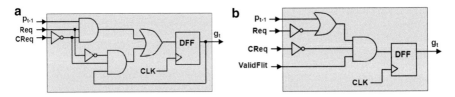

Fig. 2.26 (**a**) Conventional and (**b**) proposed grant generation circuit

Conflict Between Output Port Reservation and Flit Type

Corruption of the round-robin register results in packet loss. The input port does not
have active CReq and Req either in the middle of transferring a packet or during
no data transmission period. To differentiate these two cases, we check whether the
reservation matches to the flit type. If the current flit type is payload, there exists a
single valid reservation in one of the output ports. Missing or switched reservation
in the round-robin register can be detected and then be repaired by rewriting the
register and stopping current data flow.

Illegal Single Reservation

Because each arbiter issues a single grant each cycle, the presence of multiple
reservation signals after the RRA unit indicates miscomputing. This error is caused
by logic gate errors in round-robin computing unit. We examine the number of active
grants and reset the reservation registers if more grants exist. We assume permanent
errors on logic and storage elements have been examined and repaired in the testing
process. Stopping the grant process for one cycle, we can obtain the correct grant
signal and reservation register content in the next cycle.

2.4.4.3 Implementation

In the generic round-robin arbitration, the grant $g_{i,t}$ and priority vector $p_{i,t}$ are given
by Eqs. 2.6 and 2.7 respectively.

$$g_{i,t} = (Req \& \overline{CReq} \& p_{i,t-1}) | (\overline{Req} \& \overline{CReq} \& g_{i,t-1}) \qquad (2.6)$$
$$p_{i,t} = g_{\mathrm{mod}(i+3,4),t} | (\overline{gc_t} \& p_{i,t-1}) \qquad (2.7)$$

in which, i is the port ID, $mod(i+3,4)$ is the function to shift the priority for a five
port router without 180° transmission, t is time, gc is the bit-OR signal of all grant
signals in each output port. The circuit of conventional grant generator without error
resilience is shown in Fig. 2.26a.

   Our proposed method detects and overwrites the erroneous round-robin register
based on the appropriate operation for the given *Req*, *CReq* and *ValidFlit* infor-

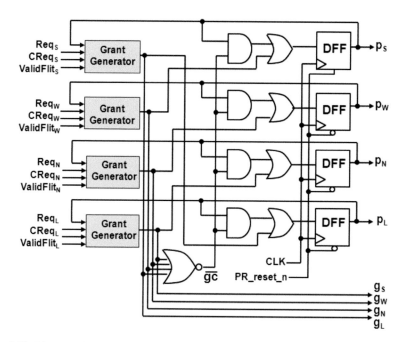

Fig. 2.27 Diagram of proposed round-robin arbitration unit

mation. The round-robin register achieves error resilience by using our new $g_{i,t}$ computation expressed in Eq. 2.8.

$$g_{i,t} = \overline{CReq} \,\& \, ValidFlit \,\& \, (p_{i,t-1}|\overline{Req}) \tag{2.8}$$

Equation 2.6 applies to our method, as well. As can be seen from Eqs. 2.6 and 2.7 the grant signal has error propagation problem: failure on $g_{i,t}$ will cause failure on $p_{i,t}$, vice versa. Thus, we propose an error termination technique for the priority register. Priority registers are reset either when the number of '1' of priority bits greater than one or when zero priority vector occurs. The reset signal PR_reset_n is given by Eq. 2.9. The reset operation may result in packet loss, but our simulation results that the impact of packet loss on the average latency and energy is significantly smaller than other approaches.

$$PR\_reset\_n = \begin{cases} 0, \; if \; \sum_{k=0}^{3} p_{i,k} \geq 2 \, or \; \sum_{k=0}^{3} p_{i,k} == 0 \\ 1, \; otherwise \end{cases} \tag{2.9}$$

The completed view of the proposed RRA unit is shown in Fig. 2.27. Note that this diagram is for the East output port. The circuit for other output ports is same with that for the east output port, except the round-robin priority. We use the proposed grant generator (shown in Fig. 2.26b) to detect and correct mute-request, multi-request and switched-request caused by errors happen to logic gates and D flipflops in the RRA unit. In addition, the signal *PR_reset_n* is used to reset the priority registers (DFFs before $p_{S/W/N/L}$). The priority registers cannot be all zeros; otherwise no request can be granted after initialization. For example, the highest priority in the east output port is for the request from the south input port.

2.4.4.4 Evaluation

Reliability

We compared the proposed method with TMR using quasi-simulation method. Because purely random simulation cannot capture each error case, we firstly verify whether the system will yield a wrong output under the different error injection conditions, to obtain $\gamma_{RRA,i,j}$ (i is the number of failed registers and j is the number of failed logic gates). The average RRA failure rate γ_{RRA} is computed by Eq. 2.10.

$$\gamma_{RRA} = \sum_{j=0}^{N_{Logic}} \sum_{i=0}^{N_{DFF}} (p_{i,j} * \gamma_{RRA\ i,j}) \tag{2.10}$$

where, the probability of having i register failures and j logic gate failures, $p_{i,j}$, is a function of the total number of register N_{DFF}, the total number of logic gates N_{Logic}, each D flipflop (DFF) failure rate ε, the failure rate ratio of logic gate over DFF β, and the specific number of failure DFFs and logic gates. The closed-form expression for $p_{i,j}$ is given by Eq. 2.11.

$$p_{i,j} = f\left(N_{DFF}, N_{Logic}, \varepsilon, \beta, i, j\right)$$
$$= \binom{i}{N_{DFF}} \varepsilon^i (1-\varepsilon)^{N_{DFF}-i} \cdot \binom{j}{N_{Logic}} (\beta\varepsilon)^j (1-\beta\varepsilon)^{N_{Logic}-j} \tag{2.11}$$

in which, $0 \le i \le N_{DFF}$, $0 \le j \le N_{Logic}$, and $\beta \le 1$.

The RRA reliability is evaluated in Fig. 2.28. Each data point for $\gamma_{RRA\ i,j}$ is obtained from the random simulation executed by 50,000 times. The values of each logic gate and D flip-flop can be flipped with the probabilities of $\beta\varepsilon$ and ε, respectively. It has been observed that storage element has higher failure rate than logic gate. However, as the circuit frequency increases, it is expected to have higher logic error rate than before. Consequently, we assume that the ratio β is no more than 1. As shown in Fig. 2.28a our method reduces the system failure rate by two orders of magnitude over the no-protection and TMR approaches for $\beta = 10^{-3}$.

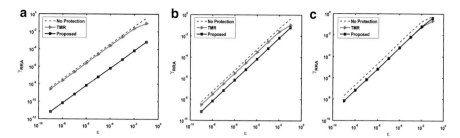

Fig. 2.28 Impact of logic gate error rate on the failure rate of round-robin arbitration unit (**a**) RRA failure rate at $\beta = 10^{-3}$ (**b**) RRA failure rate at $\beta = 10^{-1}$ (**c**) RRA failure rate at $\beta = 1$

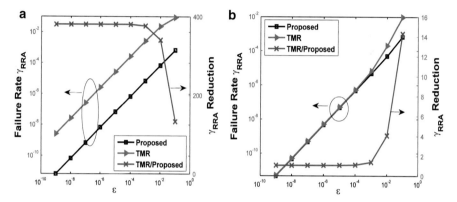

Fig. 2.29 Impact of failed flit type on the system failure rate. (**a**) Payload flit, (**b**) Tail flit for $\beta = 10^{-3}$

As β increases to 10^{-1}, our method still achieves up to 4X higher system error rate reduction than TMR, as shown in Fig. 2.28b. Our method is superior to TMR in a wide range of ε and β. When the logic gate failure rate is equal to register failure rate, the advantage of our method does not maintain, as shown in Fig. 2.28c. Figure 2.28 shows that our approach cannot reduce the failure rate to zero, partly because an incoherent state of the registers cannot be propagated through the additional control logic, eluding the detection and correction mechanisms.

In Fig. 2.29, we compare the failure rate reduction achieved by different error management methods when payload and tail flits are being transferred from input port to output port. The metric γ_{RRA} reduction is defined as the ratio of TMR failure rate over the failure rate of our method. As shown in Fig. 2.29a our method achieves up to 380X more failure rate reduction than TMR, in the case of transferring payload flits. As shown in Fig. 2.29b our method can tolerate comparable error rate to TMR method in low ε region, and the failure reduction increases to 14 times as ε increases to 10^{-2}, in the case of transferring tail flits.

Table 2.1 Area and power of routing arbitration units

	No protection	TMR	Proposed
Area (μm^2)	1982 (100%)	3822 (202%)	2160 (109%)
Dyn. Power (μW)	859.7 (100%)	2462.6 (286%)	884.3 (103%)
Leak. Power (μW)	7.2 (100%)	13.9 (193%)	8.0 (111%)

Area

Reliability improvement of TMR and our methods has been demonstrated in previous sections. Here, we compare the total area, power and delay of three different implementations for the routing arbitration stage. TMR and proposed approaches are employed to protect the RC unit and RRC unit. Error control overhead for protecting multiplex in the data path and pipeline stage registers are not included here. As shown Table 2.1, the proposed method only increases 9% area overhead over the no-protection approach; while the area overhead of TMR is more than two times to that of the no-protection approach. Compared to TMR, our method reduces area by 43%.

Power

Table 2.1 also shows the power consumption of different routing arbitration units in router stage 2. Because of fewer logic gates and D flip-flops are employed in the arbitration units, the proposed method reduces the dynamic power by 64% and the leakage power by 42%, compared to TMR. The increased total power overhead of our method over no-protection design is only 3%.

2.5 Summary

The application of Networks-on-Chip in the computing systems becomes popular, as NoC is capable to provide the scalable on-chip interconnect management. In addition to performance and energy efficiency, reliability emerges as a new challenge on the computing system design. Typical error control schemes are summarized in this chapter, and commonly used error detection/correction codes and crosstalk avoidance codes are overviewed, as well. To balance the reliability and the overhead on hardware cost and power consumption, we presented the very recent NoC link and router designs. In future work, the new error control codes and their applications in NoCs are worth more investigation to further improve the reliability for the sustainable computing systems.

References

1. Cray Research, Inc. (1985) The cray-2 computer system
2. Gioiosa R (2010) Towards sustainable exascale computing. In: Proceedings of the18th IEEE/IFIP VLSI system on chip conference (VLSI-SoC), Madrid, Spain, pp 270–275
3. Zhang Y, Sun J, Yuan G, Zhang L (2010) Perspectives of China's HPC system development: a view from the 2009 China HPC TOP100 list. J Frontiers Comput Sci China 4(4):437–444
4. Nickolls J, Dally WJ (2010) The GPU computing era. IEEE Micro 30(2):56–69
5. Truong DN et al (2009) A 167-processor computational platform in 65 nm CMOS. IEEE J Solid State Circuits 44(4):1130–1144
6. Seiler L et al (2009) Larrabee: a many-core x86 architecture for visual computing. IEEE Micro 29(1):10–21
7. Dally WJ, Towles B (2001) Route packets, not wires: on-chip interconnection networks. In: Proceedings of the 38th design automation conference (DAC'01), Las Vegas, NV, USA, pp 684–689
8. Benini L, De Micheli G (2002) Networks on chips: a new SoC paradigm. Computer 35:70–78
9. Agarwal A, Iskander C, Shankar R (2009) Survey of network on chip (NoC) architectures & contributions. Eng Comput Architec 3:1–15
10. Kogge P et al (2008) Exascale computing study: technology challenges in achieving exascale systems. Tech Rcp DARPA-2008-13, DARPA IPTO
11. Naffziger S (2006) High-performance processors in a power-limited world. In: Proceedings of the symposium on VLSI Circuits, Honolulu, Hawaii, USA, pp 93–97
12. Constantinescu C (2003) Trends and challenges in VLSI circuit reliability. IEEE Micro 23: 14–19
13. Hussein MA, He J (2005) Materials' impact on interconnect process technology and reliability. IEEE Trans Semiconduct Manuf 18:69–85
14. Jakushokas R et al (2011) Power distribution networks with on-chip decoupling capacitors. Springer, New York
15. Chandra V, Aitken R (2008) Impact of technology and voltage scaling on the soft error susceptibility in nanoscale CMOS. In: Proceedings of DFT'08, Cambridge, MA, USA, pp 114–122
16. Barsky R, Wagner IA (2004) Reliability and yield: a joint defect-oriented approach. In: Proceedings of the 19th IEEE international symposium on defect and fault tolerance in VLSI Syst (DFT'04), Cannes, France, pp 2–10
17. Shivakumar P et al (2002) Modeling the effect of technology trends on the soft error rate of combinational logic. In: Proceedings of international conference on dependable systems and networks, Washington, DC, USA, pp 389–398
18. Agarwal K, Sylvester D, Blaauw D (2006) Modeling and analysis of crosstalk noise in coupled RLC interconnects. IEEE Trans Comput Aided Des Integr Circuits Syst 25:892–901
19. Baumann R (2005) Radiation-induced soft errors in advanced semiconductor technologies. IEEE Trans Device Mater Reliab 5:305–316
20. Bertozzi D, Benini L, De Micheli G (2005) Error control scheme for on-chip communication links: the energy-reliability tradeoff. IEEE Trans Comput Aided Des Integr Circuits Syst (TCAD) 24:818–831
21. Lin S, Costello D, Miller M (1984) Automatic-repeat-request error control schemes. IEEE Commun Mag 22:5–17
22. Metzner J (1979) Improvements in block-retransmission schemes. IEEE Trans Commun COM 23:525–532
23. Lehtonen T, Lijieberg P, Plosila J (2007) Analysis of forward error correction methods for nanoscale networks-on-chip. In: Proceedings of the nano-net, Catania, Italy, pp 1–5
24. Lin S, Costello DJ (2004) Error control coding, 2nd edn. Prentice Hall
25. Sridhara S, Shanbhag RN (2005) Coding for system-on-chip networks: a unified framework. IEEE Trans Very Large Scale Integr (VLSI) Syst 12:655–667

26. Rossi D, Metra C, Nieuwland KA, Katoch A (2005) Exploiting ECC redundancy to minimize crosstalk impact. IEEE Des Test Comput 22:59–70
27. Zimmer H, Jantsch A (2003) A fault model notation and error-control scheme for switch-to-switch buses in a network-on-chip. In: Proceedings of the international conference on hardware/software codesign and system synthsis (CODES-ISSS), Newport Beach, CA, USA, pp 188–193
28. Yu Q, Ampadu P (2008) Adaptive error control for NoC switch-to-switch links in a variable noise environment. In: Proceedings of IEEE international symposiun on defect and fault tolerance in VLSI system (DFT), Cambridge, MA, USA, pp 352–360
29. Reed SI, Solomon G (1960) Polynomial codes over certain finite fields. J Soc Ind Appl Math 8:300–304
30. Dumitras T, Kerner S, Marculescu R (2003) Towards on-chip fault-tolerant communication. In: Proceedings of the Asia and South Pacific design automation conference (ASP-DAC'03), Kitakyushu, Japan, pp 225–232
31. Haas ZJ, Halpern JY, Li L (2006) Gossip-based ad hoc routing. IEEE/ACM Trans Network (TON) 14:476–491
32. Pirretti M et al (2004) Fault tolerant algorithms for network-on-chip interconnect. In: Proceedings IEEE computer society annual symposium on VLSI emerging trends in VLSI syst design (ISVLSI'04), Lafayette, Louisiana, USA, pp 46–51
33. Patooghy A, Miremadi SG (2008) LTR: a low-overhead and reliable routing algorithm for network on chips. In: Proceedings of international SoC design conference Busan, Korea, I-129–I-133
34. Bobda C et al (2005) DyNoC: a dynamic infrastructure for communication in dynamically reconfigurable devices. In: Proceedings of international conference on field programmable logic and applications, Tampere, Finland, pp 153–158
35. Zhang Z, Greiner A, Taktak S (2008) A reconfigurable routing algorithm for a fault-tolerant 2D-mesh network-on-chip. In: Proceedings of IEEE design automation conference (DAC'08), Austin, TX, USA, pp 441–446
36. Glass CJ, Ni LM (1992) The turn model for adaptive routing. In: Proceedings of international symposium computer architecture, Gold Coast, Australia, pp 278–287
37. Chiu G-M (2000) The odd-even turn model for adaptive routing. IEEE Trans Parallel Distr Syst 11:729–738
38. Li M, Zeng QA, Jone WB(2006) DyXY-A proximity congestion-aware deadlock-free dynamic routing method for network-on-chip. In: Proceedings of DAC 2006, San Francisco, CA, USA, pp 849–852
39. Hosseini A, Ragheb T, Massoud Y (2008) A fault-ware dynamic routing algorithm for on-chip networks. In: Proceedings of IEEE international symposium circuits and syst(ISCAS '08), Seattle, Washington, USA, pp 2653–2656
40. Aliabadi MR, Khademzadeh A, Raiya AM (2008) Dynamic intermediate node algorithm (DINA): a novel fault tolerance routing methodology for NoCs. In: Proceedings of international symposium on telecommunication, Tehran, Iran, pp 521–526
41. Schonwald T, Zimmermann J, Bringmann O, Rosenstiel W (2007) Fully adaptive fault-tolerant routing algorithm for network-on-chip architectures. In: Proceedings of euromicro conference on digital system design architecture, Lubeck, Germany, pp 527–534
42. Zhou J, Lau FCM (2001) Adaptive fault-tolerant wormhole routing in 2D meshes. In: Proceedings of 15th international parallel and distributed processing symposium, pp 1–8
43. Boppana RV, Chalasani S (1995) Fault-tolerant wormhole routing algorithms for mesh networks. IEEE Trans Comput 44:848–864
44. Chen K-H, Chiu G-M (1998) Fault-tolerant routing algorithm for meshes without using virtual channels. Inform Sci Eng 14:765–783
45. Park D, Nicopoulos C, Kim J, Vijaykrishnan N, Das CR (2006) Exploring fault-tolerant network-on-chip architectures. In: Proceedings of international conference on dependable syst and networks (DSN'06), Philadelphia, PA, USA, pp 93–104
46. Duato J (1997) A theory of fault-tolerant routing in wormhole networks. IEEE Trans Parallel Distr Syst 8:790–802

47. Lehtonen T, Wolpert D, Liljeberg P, Plosila J, Ampadu P (2010) Self-adaptive system for addressing permanent errors in on-chip interconnects. IEEE Trans Very Large Scale Integr (VLSI) Syst 18:527–540
48. Lehtonen T, Liljeberg P, Plosila J (2007) Online reconfigurable self-timed links for fault tolerant NoC. VLSI Des 2007:1–13
49. Elias P (1954) Error-free coding. IEEE Trans Inf Theory 4:29–37
50. Fujiwara E (2006) Code design for dependable systems: theory and practical applications. Wiley Interscience, Hoboken
51. Pyndiah R (1998) Near-optimum decoding of product codes: block turbo codes. IEEE Trans Commun 46(8):1003–1010
52. Fu B, Ampadu P (2009) On hamming product codes with type-II hybrid ARQ for on-chip interconnects. IEEE Trans Circuits Syst I, Reg Papers 9:2042–2054
53. Constantinides K et al (2006) BulletProof: a defect-tolerant CMP switch architecture. In: Proceedings of HPCA'06, Austin, Feb 2006, pp 5–16
54. Patel KN, Markov IL (2004) Error-correction and crosstalk avoidance in DSM busses. IEEE Trans Very Large Scale Integr (VLSI) Syst 12:1076–1080
55. Ganguly A, Pande PP, Belzer B, Grecu C (2008) Design of low power & reliable networks on chip through joint crosstalk avoidance and multiple error correction coding. J Electron Test Theory Appl (JETTA), Special Issue on Defect and Fault Tolerance 24:67–81
56. Ganguly A, Pande PP, Belzer B (2009) Crosstalk-aware channel coding schemes for energy efficient and reliable NOC interconnects. IEEE Trans Very Large Scale Integr (VLSI) Syst 17(11):1626–1639
57. Sridhara S, Shanbhag RN (2007) Coding for reliable on-chip buses: a class of fundamental bounds and practical codes. IEEE Trans Comput Aided Des Integr Circuits Syst 5:977–982
58. Sridhara S, Ahmed A, Shanbhag RN (2004) Area and energy-efficient crosstalk avoidance codes for on-chip busses. In: Proceedings of international conference on computer design (ICCD), San Jose, CA, USA, pp 12–17
59. Duan C, Tirumala A, Khatri SP (2001) Analysis and avoidance of crosstalk in on-chip buses. In: Proceedings of hot interconnects, Stanford, California, USA, pp 133–138
60. Victor B, Keutzer K (2001) Bus encoding to prevent crosstalk delay. In: Proceedings of IEEE/ACM international conference on computer-aided design (ICCAD), San Jose, CA, USA, pp 57–63
61. Hirose K, Yassura H (2000) A bus delay reduction technique considering crosstalk. In: Proceedings of design, automation and test in Europe (DATE), Paris, France, pp 441–445
62. Nose K, Sakurai T (2001) Two schemes to reduce interconnect delay in bi-directional and uni-directional buses. In: Proceedings of VLSI symposium, Kyoto, Japan, pp 193–194
63. Fu B, Ampadu P (2010) Exploiting parity computation latency for on-chip crosstalk reduction. IEEE Trans Circuits Syst II: Expr Briefs 57:399–403
64. Arizona State University Predictive Technology Model [Online]. http://ptm.asu.edu/
65. Fick D et al. (2009) A highly resilient routing algorithm for fault-tolerant NoCs. In: Proceedings of DATE'09, Nice, France, Mar 2009, pp 21–26
66. Sanusi A, Bayoumi MA (2009) Smart-flooding: a novel scheme for fault-tolerant NoCs. In: Proceedings of IEEE SoC conference, Belfast, Northern Ireland, Sept 2009, pp 259–262
67. Rodrigo S, Flich J, Roca A, Medardoni S, Bertozzi D, Camacho J, Silla F, Duato J (2010) Addressing manufacturing challenges with cost-efficient fault tolerant routing. In: Proceedings of NOCS'10, Grenoble, France, May 2010, pp 25–32
68. Yanamandra A et al (2010) Optimizing power and performance for reliable on-chip networks. In: Proceedings of ASP-DAC'10, Taipei, Taiwan, Jan 2010, pp 431–436
69. Lyons REAND, Vanderkulk W (1962) The use of triple-modular redundancy to improve computer reliability. IBM J Res Dev 6(2):200–209
70. Vangal S et al (2008) An 80-tile sub-100-W TeraFLOPS processor in 65-nm CMOS. IEEE J Solid State Circuits 43(1):29–41
71. Yu Q, Zhang M, Ampadu P (2011) Exploiting inherent information redundancy to manage transient errors in NoC routing arbitration. In: Proceedings of. 5th ACM/IEEE international symposium on networks-on-chip (NoCS'11), Pittsburgh, Pennsylvania, USA, pp 105–112

Chapter 3
Energy Adaptive Computing for a Sustainable ICT Ecosystem

Krishna Kant, Muthukumar Murugan, and David Hung Chang Du

3.1 Introduction

Information and Computing Technology (ICT) has traditionally emphasized primarily on the performance of both hardware and software. Lately power/thermal issues of ICT equipment have forced a consideration of these aspects on par with performance. Power/thermal issues arise at all levels from transistors up to entire ecosystems that involve data centers, clients and the intervening network. At the architectural level, the increasing speeds, smaller feature sizes, and exploding wire widths (and hence resistance) all conspire to make power/thermal issues the main architectural hurdle in sustaining Moore's law. At higher levels, the smaller form factors and more difficult cooling aggravate the problem further. Although power/thermal management is an active area of research, power/thermal issues are still largely approached in the form of opportunistic methods to reduce power consumption or stay within the thermal profile while minimizing any performance impact [1, 2]. Much of this research is focused on reducing the direct energy usage of the data center, whereas from an environment impact perspective one needs to consider the entire life-cycle of energy consumption – that is, the energy consumption in the manufacture, distribution, installation, operation and disposal of the entire data center infrastructure including IT assets, power distribution equipment, and cooling infrastructure [3].

K. Kant (✉)
George Mason University, Fairfax, VA, USA
e-mail: kkant@gmu.edu

M. Murugan • D.H.C. Du
University of Minnesota, Minneapolis, MN, USA
e-mail: murugan@cs.umn.edu; du@cs.umn.edu

P.P. Pande et al. (eds.), *Design Technologies for Green and Sustainable Computing Systems*, 59
DOI 10.1007/978-1-4614-4975-1_3, © Springer Science+Business Media New York 2013

Looking at energy consumption from this larger perspective entails not only low power consumption during operation but also leaner designs and operation using renewable energy as far as possible. Thus, the fundamental paradigm that we consider is to replace the traditional overdesign at all levels with rightsizing coupled with smart control in order to address the inevitable lack of capacity that may arise occasionally. In general, such lack of capacity may apply to any resource; however, we only consider its manifestation in terms of energy/power constraints. Note that power constraints could relate to both real constraints in the power availability as well as the inability to consume full power due to cooling/thermal limitations. Power consumption limitation indirectly relates to capacity limitation of other resources as well, particularly the dominant ones such as CPU, memory, and secondary storage devices. We call this as *energy adaptive computing* or EAC [4, 5].[1] The main point of EAC is to consider energy related constraints at all levels and dynamically adapt the computation to it as far as possible. A direct use of locally produced renewable energy could reduce the distribution infrastructure, but must cope with its often variable nature. Thus, better adaptation mechanisms allow for more direct use of renewable energy.

It is well recognized by now that much of the power consumed by a data center is either wasted or used for purposes other than computing. In particular, when not managed properly, up to 50% of the data center power may be used for purposes such as chilling plant operation, compressors, air movement (fans), electrical conversion and distribution, and lighting [6]. It follows that from a sustainability perspective, it is not enough to simply minimize operational energy usage or wastage; we need to minimize the energy that goes into the infrastructure as well. This principle applies not only to the supporting infrastructure but to the IT devices such as clients and servers themselves. Even for servers in data centers, the increased emphasis on reducing operating energy makes the non-operational part of the energy a larger percentage of the life-cycle energy consumption and could almost account for 50% [3]. For the rapidly proliferating small mobile clients such as cell-phones and PDAs, the energy used in their manufacture, distribution and recycling could be a dominant part of the life-time energy consumption. Towards this end, it is important to consider data centers that can be operated directly via locally produced renewable energy (wind, solar, geothermal, etc.) with minimal dependence on the power grid or large energy storage systems. Such an approach reduces carbon footprint not only via the use of renewable energy but also by reducing the size and capacity of power storage and power-grid related infrastructure. For example, a lower power draw from the grid would require less heavy-duty power conversion infrastructure and reduce its cost and carbon footprint. The down-side of the approach is more variable energy supply and more frequent episodes of inadequate available energy to which the data center needs to adapt

---

[1] Here "energy adaptation" implicitly includes power and thermal adaptation as well.

dynamically. Although this issue can be addressed via large energy storage capacity, energy storage is currently very expensive and would increase the energy footprint of the infrastructure.

The power and cooling infrastructure in servers, chassis, racks, and the entire data center is designed for worst-case scenarios which are either rare or do not even occur in realistic environments. We argue for much leaner design of all components having to do with power/thermal issues: heat sinks, power supplies, fans, voltage regulators, power supply capacitors, power distribution network, Uninterrupted Power Supply (UPS), air conditioning equipment, etc. This leanness of the infrastructure could be either static (e.g., lower capacity power supplies and heat sinks, smaller disks, DRAM, etc.), or dynamic (e.g., phase shedding power supplies, hardware resources dynamically shared via virtualization). In either case, it is necessary to adapt computations to the limits imposed by power and thermal considerations. We assume that in all cases the design is such that limits are exceeded only occasionally, not routinely.

3.2   Challenges in Energy Adaptive Computing

It is clear from the above discussion that many advanced techniques for improving energy efficiency of IT infrastructure and making it more sustainable involve the need to dynamically adapt computation to the suitable energy profile. In some cases, this energy profile may be dictated by energy (or power) availability, in other cases the limitation may be a result of thermal/cooling constraints. In many cases, the performance and/or QoS requirements are malleable and can be exploited for energy adaptation. For example, under energy challenged situations, a user may be willing to accept longer response times, lower audio/video quality, less up to date information, and even less accurate results. These aspects have been explored extensively in specific contexts, such as adaptation of mobile clients to intelligently manage battery lifetime [7]. However, complex distributed computing environments provide a variety of opportunities for coordinated adaptation among multiple nodes and at multiple levels [8]. In general, there are three types of distributed energy adaptation scenarios: (a) Cluster computing (or server to server), (b) Client-server, and (c) Peer to Peer (or client to client). These are shown pictorially in Fig. 3.1 using dashed ovals for the included components and are discussed briefly in the following. Notice that in all cases, the network and the storage infrastructure (not shown) are also important components that we need to consider in the adaptation.

Although we discuss these three scenarios separately, they generally need to be addressed together because of multiple applications and interactions between them. For example, a data center would typically support both client-server and cluster applications simultaneously. Similarly, a client may be simultaneously involved in both peer-to-peer and client-server applications.

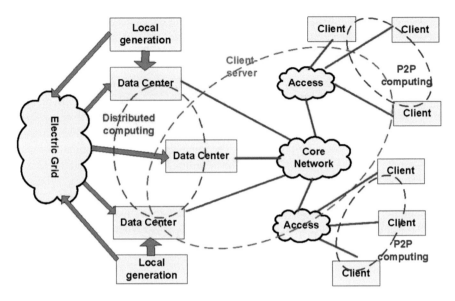

Fig. 3.1 Illustration of energy adaptation loops

3.2.1   Challenges in a Cluster Environment

Cluster EAC refers to computational models where the request submitted by a client requires significant computation involving multiple servers before the response can be returned. That is, client involvement in the service is rather minimal, and the energy adaptation primarily concerns the data center infrastructure. In particular, a significant portion of the power consumed may go into the storage and data center network and they must be considered in adaptation in addition to the servers themselves.

In cluster EAC, the energy adaptation must happen at multiple levels such as subsystems within a server, servers in a rack, etc. At each level there may be a power limit that the level must adapt to. Some of the limits may be "soft" in the sense that they simply represent algorithmic allocation of available energy, and intelligent estimation and adjustment of these limits is crucial. At the highest level, energy adaptation is required to conform to the power generation (or supply) profile of the energy infrastructure. Power limits may also need to be established for different types of infrastructure, for example, the compute, storage and network portions of the infrastructure.

In addition to the available energy, the thermal constraints play a significant role in workload adaptation. Traditionally, CPUs are the only devices that have significant thermal issues to provide both thermal sensors and thermal throttling mechanisms to ensure that the temperature stays within appropriate limits. For example, the T states provided by contemporary CPUs allows introduction of dead cycles periodically in order to let the cores cool. DIMMs are also beginning to be fitted with thermal sensors along with mechanisms to reduce the heat load. With tight enclosures such as blade servers and laptop PCs, ambient cooling, and increasing power consumption, other components (e.g. switching fabrics, interconnects, shared cache, etc.) are also likely to experience thermal issues. In challenging thermal environments, a coordinated thermal management is crucial because the consequences of violating a thermal limit could be quite severe. Also, an over throttling of power to provide a conservative temperature control could have severe performance implications.

Thermal control at the system level is driven by cooling characteristics. For example, it is often observed that all servers in a rack do not receive the same degree of cooling, instead, depending on the location of cooling vents and air movement patterns, certain servers may receive better cooling than others. Most data centers are unlikely to have finer grain mechanisms (e.g., air direction flaps) to even out the cooling effectiveness. Instead, it is much easier to do their thermal management to conform to the cooling profile. So, the simplest scheme is for each server to manage its own thermals based on the prevailing conditions (e.g., on-board temperature measurements). However, such independent controls can lead to unstable or suboptimal control. A coordinated approach such as the one considered in [5] could be used to ensure satisfactory operation while staying within the temperature limits or rather within the power limits dictated by the temperature limit and heat dissipation characteristics.

3.2.1.1   Estimation and Allocation of Energy

An important aspect of managing a resource is the ability to easily measure resource consumption of the desired software component (e.g., a task, VM, or application) while it is running and accurately estimate the resource requirements before the software is run. The purpose of the latter is to decide where, when, and how to run the software. Unfortunately, when the resource in question is energy (or power), both the measurement or estimation can be quite difficult. Part of the difficulty in measurement arises from the fact that the direct power measurement capability is often unavailable, and the power consumption must be estimated indirectly via available counters in the platform. For example, a direct measurement of power consumption of an individual CPU core is often not feasible and a standard method is to compute power based on a variety of low-level performance monitoring counters that are available on-chip. An even more difficult issue is to break the power consumption of a physical entity down to the software entities (e.g., VMs or tasks) using it [9].

An a-priori estimation of power consumption is difficult because the energy consumption not only depends on workload and hardware configuration but also on complex interactions between various hardware and software components and power management actions. For example, energy consumed by the CPU depends on the misses in the cache hierarchy, type of instructions executed, and many other micro-architectural details and how they relate to the workload being executed. Furthermore, when multiple software components are running together on the same hardware, they can interact in complex ways (e.g., cache working set of one task affected by presence of another task). Consequently, neither the performance nor the power consumption adds up linearly, e.g., the active power for two VMs running together on a server does not equal the sum of active powers of individual VMs on the same server. Thus accurate energy estimation remains a difficult problem that we do not tackle in this article.

3.2.1.2   Planning and Execution of Control Actions

Given the power and QoS constraints, the first step in any power control mechanism is to design an optimal/close to optimal solution that would achieve the target performance. However these control actions are not instantaneous and involve overheads. Realizing these solutions in real time involves multiple steps that include time consuming operations like switching a server from sleep/low power modes to fully operational modes or vice versa. The latency involved in these operations is significant and cannot be overlooked. The state changes involved may cause transient instabilities in the participating components. For instance in a datacenter environment, when a server is shut down, the load it was handling needs to be redistributed to other servers. This sudden increase in load in the other servers causes the applications already running on them to slow down. Also, the control actions themselves consume some resources for their execution. All these factors call for a careful planning and execution of the power control actions.

The initiation of the planning process is based on certain events e.g., decrease in available power supply, increase in application traffic etc. If these processes are reactive, i.e., they are initiated after the event has occurred, the associated delays will be extremely large. Hence the events need to be predicted and the necessary control actions need to be initiated beforehand. The planning process can also be made more dynamic by means of Model Predictive Control (MPC) [10] techniques where the actions are planned for every time instant $t + \tau$, $\forall \tau \in \{0, 1, ..T\}$, at time t, and only the control action for time $t + 1$ is implemented. The same process is repeated for the rest of the receding time horizon T. However these techniques could turn out to be expensive when the state space is large. Also these techniques might take a long time to converge and are not essentially optimal.

3.2.2   Challenges in Other Environments

In this section, we briefly address the client-server and peer-to-peer (P2P) environments. In its full generality, client-server EAC needs to deal with a coordinated end-to-end adaptation including the client, server, and the intervening network. The purpose of the coordination is to optimize the client experience within the energy constraints of each of the three components. As the clients become more mobile and demand richer capabilities, the limited battery capacity gets in the way and the energy adaptation can help. Furthermore, since these devices are also constrained in terms of their compute power, energy and network bandwidth, the adaptation goes well beyond just the energy. For example, techniques have been proposed to outsource mobile computation to the cloud platforms that can provide the required resources on demand [11–13]. In particular, complex and compute intensive image processing tasks can be migrated from the mobile clients to the cloud, and this could be considered as a broader energy adaptation since the migration does conserve battery life. Adding energy adaptation on the server side to these mechanisms makes them particularly difficult to handle. One interesting approach is to adapt the allocation of resources on the server side based on the remaining energy of the mobile clients.

Client-server EAC can be supported by defining client energy states and the QoS that the client is willing to tolerate in different energy states as a contract and then do a contract adaptation, as in [14]. However, since server-side adaptation (such as putting the server in deep sleep state and migrating the application to another server) can affect many clients, the client contracts play a role in where the client applications are hosted on servers and how the servers themselves adapt. This coupling between client and servers, along with appropriate network power management actions makes the overall coordination problem very difficult.

Energy adaptation in P2P environment requires cooperation among peers. This issue is examined in [15]. The authors propose an energy adaptive version of the Bit Torrent protocol. The battery constrained clients define an energy budget for downloading the file which enables them to adapt their contributions to the network and the service they receive from the network based on it. The protocol ensures the provisioning and delivery of the desired service rate to the clients based on their energy budget. The proposed mechanism exploits the following two characteristics of the mobile devices to adapt their energy consumption based on their energy budgets.

1. A mobile device consumes more energy when transmitting than receiving. Essentially, during transmission, the signal is amplified to achieve the desired signal to noise ratio for successful decoding at the receiver. Thus attributing for the higher energy consumption.
2. It is much more energy efficient for the mobile device to download a file faster, i.e., at high download rate.

After each transmission or reception, mobile devices continue to remain in active state for a short duration, called tail time, in anticipation of another packet. Frequent occurrence of tail time can result in significant energy consumption for the mobile devices [16]. At high download rates, packets are either received in the tail time or in large single bursts, thus, preventing the frequent occurrence of tail time and reducing the average energy per transfer [16].

3.3   Realizing Energy Adaptive Computing in Datacenters

In this section we discuss two specific cases of EAC in a cluster environment in a datacenter.

1. The first case is a datacenter with transactional applications where clients send queries to the applications running on servers in the datacenter and the response is sent back to the clients. We assume that the applications are running in virtual machines that can be migrated between servers. We discuss the design and implementation details of a controller called *Willow* that reacts to changes in supply and demand side migrations and migrates work from energy deficient to energy surplus regions.
2. The second case also considers transactional workloads. However, the client sessions are long lived and each query is processed by multiple tiers of servers and finally the response is sent back to the client. We discuss strategies for planning the control actions for energy adaptations in such a scenario.

3.3.1   Willow: Controller for Energy and Thermal Adaptive Computing in Datacenters

This section describes in detail the design and implementation of a control system named *Willow* for energy and thermal adaptive computing in datacenters.

3.3.1.1   Hierarchical Power Control

Power/energy management is often required at multiple levels including individual devices (CPU cores, memory DIMMs, NICs, etc.), subsystems (e.g., CPU - cache subsystem), systems (e.g., entire servers), and groups of systems (e.g., chassis or racks). In a power limited situation, each level will be expected to have its own power budget, which gets divided up into power budgets for the components at the next level. This brings in extra complexity since one must consider both the demand and supply sides in a coordinated fashion at various levels. One simple such multi-level power management model is shown in Fig. 3.2. The data center level power management unit (PMU) is at the level 3. The rack level PMU is at level 2 and server/switch level PMUs are at level 1. With such a multilevel power management architecture our control scheme attempts to provide the scalability required for handling energy and thermal adaptation in large data centers with minimum impact on the underlying networks.

In the hierarchical power control model that we have assumed, the power budget in every level gets distributed to its children nodes in proportion to their demands. All the leaf nodes are in level 0. The component in each level $l+1$ has configuration

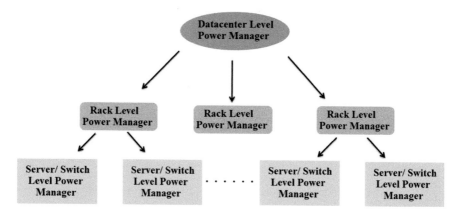

Fig. 3.2 Hierarchical multi-level power control in a datacenter

information about the children nodes in level l. For example the rack level power manager has to have knowledge of the power and thermal characteristics of the individual components in the rack. The components at level l continuously monitor the demands and utilization levels and report them to level $l+1$. This helps level $l+1$ to continuously adjust the power budgets. Level $l + 1$ then directs the components in level l as to what control action needs to be taken. The granularities at which the monitoring of power usage and the allocation adjustments are done are different and are discussed in detail later.

3.3.1.2   Energy-Temperature Relationship

In the design of our control scheme we limit the power consumption of a device based on its thermal limits as follows.

Let t denote time, $T(t)$ the temperature of the component as a function of time, $P(t)$ power consumption as a function of time, and c_1, c_2 be the appropriate thermal constants. Also, let T_a denote the ambient temperature, i.e., temperature of the medium right outside the component. The component will eventually achieve this temperature if no power is supplied to it. Then the rate of change of temperature is given by

$$dT(t) = [c_1 P(t) + c_2(T(t) - T_a)]dt \qquad (3.1)$$

Being a first-order linear differential equation, this equation has an explicit solution. Let $T(0)$ denote the temperature at time $t = 0$. Then,

$$T(t) = [T_a + [T(0) - T_a]e^{-c_2 t}] + c_1 e^{-c_2 t} \int_0^t P(\tau)e^{c_2 \tau} \, d\tau \qquad (3.2)$$

where the first term relates to cooling and tends to the ambient temperature T_a and the second term relates to heating. Let T_{limit} denote the limit on the temperature and P_{limit} is the limit on power consumption so that the temperature does not exceed T_{limit} during the next adjustment window of Δ_s seconds. It is easy to see that,

$$T(\tau) = T_a + P_{limit}c_1/c_2[1 - e^{-c_2\Delta_s}] + [T(0) - T_a]e^{-c_2\Delta_s} \qquad (3.3)$$

It can be observed that Eq. 3.2 can be used to predict the value of temperature of the device at the end of the next adjustment window and hence can help in making the migration decisions. We use this relationship to estimate the maximum power consumption that can be allowed on a node so that it does not exceed its thermal limits.

3.3.1.3 Time Granularity

The utilization of server resources in a data center varies widely over a large scale. If the nature of the workload fluctuates significantly, it is likely that different resources (e.g., CPU cores, DRAM, memory bus, platform links, CPU core interconnects, I/O adapters, etc.) become bottlenecks at different times; however, for a workload with stable characteristics (but possibly varying intensity) and a well-apportioned server, there is one resource (typically CPU and sometimes network adapter) that becomes the first bottleneck and its utilization can be referred to as server utilization. We assume this is the case for our modeling presented in this article, since it is extremely difficult to deal with arbitrarily configured servers running workloads that vary not only in intensity but their nature as well. Under these assumptions, the power consumption can be assumed to be a linear monotonic function of the utilization.

Because of varying intensity of the workload, it is important to deal with average utilizations of the server at a suitable time granularity. For convenience the demand side adaptations are discretized with a time granularity of Δ_{Dl}. It is assumed that this time granularity is sufficiently coarse to accommodate accurate power measurement and its presentation, which can be quite slow. Typically, appropriate time granularity at the level of individual servers are of the order of tens of milliseconds or more. Coarser granularities may be required at higher levels (such as rack level).

Even with a suitable choice of Δ_{Dl}, it may be necessary to do further smoothing in order to determine trend in power consumption. Let $CP_{l,i}$ be the power demand of node i at level l. For exponential smoothing with parameter $0 < \alpha < 1$, the smoothed power demand CP' is given by:

$$CP'_{l,i} = \alpha CP_{l,i} + (1 - \alpha)CP'^{old}_{l,i} \qquad (3.4)$$

Note that the considerations in setting up the value of Δ_{Dl} come from the demand side. In contrast, the supply side time constants are typically much larger. Because of the presence of battery backed UPS and other energy storage devices, any temporary deficit in power supply in a data center is integrated out. Hence the supply side time constants are assumed to be $\Delta_{Sl} = \eta_1 \Delta_{Dl}$, where η_1 is an integer > 1. *Willow* also performs workload consolidation when the demand in a server is very low so that some servers can be put in a deep sleep state such as S3 (suspend to memory) or even S4 (suspend to disk). Since the activation/deactivation latency for these sleep modes can be quite high, we use another time constant Δ_{Al} for making consolidation related decisions. We assume $\Delta_{Al} = \eta_2 \Delta_{Dl}$, for some integer η_2 such that $\eta_2 > \eta_1$.

3.3.1.4 Supply Side Adaptation

As mentioned earlier we ignore the case where the data center operates in a perpetually energy deficient regime. The available power budget of any level $l + 1$ is allocated among the nodes in level l proportional to their demands. As mentioned before, the supply side adaptations are done at a time granularity of Δ_{Sl}. Hence the power budget changes are reflected at the end of every Δ_{Sl} time period. Let TP_{l+1}^{old} be the overall power budget at level $l + 1$ during the last period. TP_{l+1} is the overall power budget at the end of current period. $\Delta_{TP} = \text{TP}_{l+1} - \text{TP}_{l+1}^{old}$ is the change in overall power budget. If Δ_{TP} is small we can update the values of $\text{TP}_{l,i}$'s rather trivially. However if Δ_{TP} is large we need to reallocate the power budgets of nodes in level l. In doing so we consider both *hard* constraints due to power limitations of devices and *soft* constraints due to available power budgets.

The power and thermal constraints thus necessitate the migration of demand in level l from power deficient nodes to nodes with surplus power budget. Any increase in the overall power budget happens at a higher level and is then reflected in its constituent lower levels. This situation can lead to three subsequent actions.

1. If there are any under provisioned nodes they are allocated just enough power budget to satisfy their demand.
2. The available surplus can be harnessed by bringing in additional workload.
3. If surplus is still available at a node then the surplus budget is allocated to its children nodes proportional to their demand.

3.3.1.5 Demand Side Adaptation

The demand side adaptation to thermal and energy profiles is done systematically via migrations of the demands. We assume that the fine grained power control in individual nodes is already being done so that any available idle power savings can be harvested. Our focus is on workload migration strategies to adapt to the energy deficient situations. For specificity we consider only those type of applications in which the demand is driven by user queries and there is minimum or no interaction between servers, (e.g.,) transactional workloads. The applications are hosted by one or more virtual machines (VMs) and the demand is migrated between nodes by migrating these virtual machines. Hence the power consumption is controlled by simply directing the user queries to the appropriate servers hosting them.

We carefully avoid pitfalls like oscillations in decisions by allowing sufficient margins both at the source and the destination to accommodate fluctuations after the migrations are done. The migrations are initiated in a bottom up manner. If the power budget $TP_{l,i}$ of any component i is too small then some of the workload is migrated to one of its sibling nodes. We call this as local migration. Only when local migrations to sibling nodes is not possible non-local migrations are done.

The migration decisions are made in a distributed manner at each level in the hierarchy starting from the lowermost level. The local demands are first satisfied with the local surpluses and then those demands that are not satisfied locally are passed up the hierarchy to be satisfied non-locally. Now we define a few terms related to the migration decisions.

Power Deficit and Surplus: The power deficit and surplus of a component i at level l are defined as follows.

$$P_{def}(l,i) = [CP'_{l,i} - TP_{l,i}]^+ \tag{3.5}$$

$$P_{sur}(l,i) = [TP_{l,i} - CP'_{l,i}]^+ \tag{3.6}$$

where $[]^+$ means if the difference is negative it is considered zero.

If there is no surplus that can satisfy the deficit in a node, the excess demand is simply dropped. In practice this means that some of the applications that are hosted in the node are either shut down completely or run in a degraded operational mode to stay within the power budget.

Power Margin (P_{min}): The minimum amount of surplus that has to be present after a migration in both the source and target nodes of the migration. This helps in mitigating the effects of fluctuations in the demands.

Migration Cost: The migration cost is a measure of the amount of work done in the source and target nodes of the migrations as well as in the switches involved in the migrations. This cost is added as a temporary power demand to the nodes involved.

A migration is done if and only if the source and target nodes can have a surplus of at least P_{min}. Also migrations are done at the application level and hence the demand is not split between multiple nodes. Finally *Willow* also does resource consolidation to save power whenever possible. When the utilization in a node is really small the demand from that node is migrated away from it and the node is deactivated.

The matching of power deficits to surpluses is done by a variable sized bin packing algorithm called FFDLR [17] solves a bin packing problem of size n in time $O(n \log n)$. The optimality bound guaranteed for the solution is $(3/2) OPT + 1$ where OPT is the solution given by an optimal bin packing strategy.

Willow implements a unidirectional hierarchical power control scheme. Migrations of power demands are initiated by the power and thermal constraints introduced as a result of increase in demand at a particular node or decrease in power budget to the node. Simultaneous supply and demand side adaptations are done to

Table 3.1 Utilization vs. power consumption

Application class	SLA requirement
Type I	Average delay $\leq$ 120 ms, cannot be migrated
Type II	Average delay $\leq$ 180 ms, can be migrated
Type III	Average delay $\leq$ 200 ms, can be migrated

match the demands and power budgets of the components. In the next section we evaluate *Willow* via detailed experiments and simulations.

3.3.1.6 Assumptions and QoS Model

We built a simulator in Java for evaluating the ability of our control scheme to cater to the QoS requirements of tasks when there are energy variations. The Java simulator can be configured to simulate any number of nodes and levels in the hierarchy. For our evaluations we used the configuration with 18 nodes and 3 types of applications. The application types and their SLA requirements are shown in Table 3.1. For simplicity we assume that each application is hosted in a VM and can be run on any of the 18 nodes. To begin with, each node is assigned a random mix of applications. The static power consumption of nodes is assumed to be 20% of the maximum power limit (450 W). There is a fixed power cost associated with migrations. In the configuration that we use for our experiments, we assume a single SAN storage that can be accessed by all nodes in the data center. Storage migration is done when the VM disk files have to be migrated across shared storage arrays due to shortage of storage capacity and is usually dealt with separately (e.g., Storage VMotion [18]) so as to reduce the delays involved. Since we deal with compute intensive applications in our experiments, we assume a single shared storage domain is large enough to support the applications and we do not account for delays involving data migrations.

Since our experimental platform consists of multiple VMs with independent traffic patterns, initially we ran our experiments for the case where the traffic to each individual VM was a Poisson process. In order to test our proposed scheme with real world traces, we used the Soccer World Cup 98 [19] traces for our evaluation. In this section, we present only the results with the World Cup 98 trace. The trace dataset consists of all the requests made to the 1998 Soccer World Cup Web site during April to July, 1998. Since the trace data was collected from multiple physical servers, we had to process the traces before we could use them in our virtualized environment. We used traces collected on different days for different VMs. We scaled the arrival rates of queries to achieve the target utilization levels. We assume that the queries in the traces belonged to the three classes of applications as shown in Table 3.1. The service times for queries for each application class is different (10, 15 and 20 ms respectively) and the energy consumed by a query is assumed to be proportional to its runtime. The time constant multipliers for discrete time control

η_1 and η_2 are assumed to be 4 and 7 respectively. Unless specified otherwise the ambient temperature of the nodes was assumed to be 25 °C. Also the thermal limit of the servers and switches is assumed to be 70 °C. The thermal constants in Eq. 3.1 were determined to be $c_1 = 0.2$ $c_2 = -0.008$ from the experiments as described in [5]. We use these values for our simulations as well.

We measure the utilization of a VM based on CPU time spent by the VM in servicing the requests. However the actual utilization may be higher. For instance the actual utilization of a task may be increased due to CPU stalls that are caused by memory contention from other tasks or context switches between multiple VMs. In our simulations we include a factor called the interference penalty for the utilization and power/energy calculations. Basically the idea is that when n tasks are running on a server, the utilization of each task is increased due to the interference from the other $(n - 1)$ tasks. Hence the actual utilization of an application is calculated as follows.

$$U_i = U_i + \alpha \sum_j U_j, \forall j \in \{1, 2, \ldots n\} - \{i\}$$
$$U_{node} = \min [1.0, \sum_{i=1}^{n} U_i]$$

where n is the total number of applications in the node
U_i is the utilization of ith application, $i \in \{1, 2, \ldots n\}$.
U_{node} is the actual utilization of the node.

We then calculate the average power consumed in the node based on the actual average utilization of the node. We conducted a few experiments on a Dell machine running VMWare ESX server to determine the value of α. We varied the number of VMs and their utilization levels and compared the sum of their utilizations with the actual utilization reported by the ESX server. We then calculated the value of α using simple first order linear regression. The value of α was found to be 0.01. Note that the workload that we used in our experiments was totally CPU bound. For other workloads that involve memory or network contention, the interference penalty might be higher.

3.3.1.7 QoS Aware Scheduler

Traditional scheduling algorithms like round robin and priority based scheduling are not QoS aware and do not have any feedback mechanism to guarantee the QoS requirements of jobs. A few algorithms that attempt to guarantee some level of QoS do so by mechanisms like priority treatment and admission control. Our objective in this work is to allow for energy adaptation while respecting QoS needs of various applications to the maximum extent possible. In this regard we implemented a QoS aware scheduling algorithm in the nodes as shown in Fig. 3.3. The scheduler uses an integral controller to adjust the weights of the applications at regular intervals based on the delay violations of the applications. The applications are allocated CPU shares proportional to their weights. Applications with higher delay violations get more CPU share. It is well known from classic queuing theory [20] that in an asymptotic sense, as $U \to 1$ the wait time of jobs is directly proportional to $1/(1-U)$

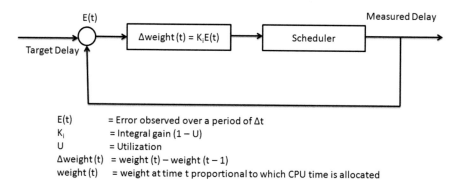

E(t)   = Error observed over a period of Δt
K$_i$ = Integral gain (1 – U)
U   = Utilization
Δweight (t) = weight (t) – weight (t – 1)
weight (t) = weight at time t proportional to which CPU time is allocated

Fig. 3.3 QoS aware scheduler

Time Interval : Δ_{AI}	Consolidate	• Shut down nodes that have very low utilization
Time Interval : Δ_{SI}	Supply Side Adaptations	• Adjust power allocations (respect thermal limits and demand)
Time Interval : Δ_{DI}	Demand Side Adaptations	• Predict Demand (Eq 5) • Migrations (Bin Packing)
Time Interval : Δ_t	Scheduler (Integral Controller)	

Fig. 3.4 Various adaptations done in *Willow* and the different time granularities

where U is the overall queue utilization. Hence the integral gain for the controller is calculated as $(1 - U)$. Using the overall utilization as the integral gain also avoids oscillations and keeps the system stable. The error in delay $E(t)$ for each application is the difference between the delay bound and the measured delay. At the end of each sample interval t (of size Δt), the new weights are calculated as follows.

$$weight(t) = weight(t-1) + K_i * E(t) \qquad (3.7)$$

Then the CPU shares are allocated to the applications proportional to their weights.

Figure 3.4 shows an overall picture of the different adaptation mechanisms that are done in *Willow* at different time granularities. The scheduler works in the individual nodes at the smallest time granularity. At the next higher time granularity the demand side adaptations are done that include migration of deficits

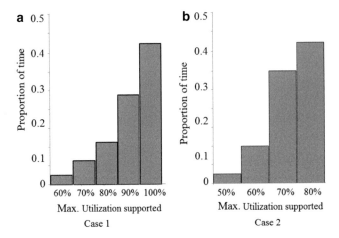

Fig. 3.5 Power supply profiles used – CDF of the total available power budgets as a function of the maximum utilization levels supported

3.3.1.8 Experimental Results

We use the response time as a metric to quantify the impact of EAC in the presence of variations in energy. The response time includes the time that the queries wait in the queue to be scheduled and the run time of the queries. We show that the response times are improved when adaptations to the energy variations are done in *Willow*. The significance of *Willow* is realized the most when the devices are operating at the edge – that is when the power budgets are just enough to meet the aggregate demand in the data center and there is no over provisioning. To demonstrate this, we tested the performance of *Willow* with two different cases of power budgets as shown in Fig. 3.5. Let P_U be the power required to support the data center operations when the average utilization of the servers is $U\%$. Figure 3.5 shows the cumulative distribution of the total available power budget values during the simulation of 350 min and the proportion of time for which the particular power budget value was available. The first case (Case 1) is when the total available power budget varies between P_{100} and P_{60}. The second case (Case 2) is when the total power budget varies between P_{80} and P_{50}.

Figure 3.6a compares the percentage of queries with delay violations in Case 1 when the QoS aware scheduler described in Sect. 3.3.1.7 is used alone and when the scheduler is used in combination with *Willow*. We see that at low utilizations the

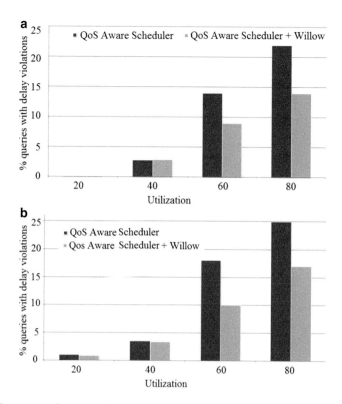

Fig. 3.6 Percentage of queries with delay violations when the power profile used is as shown in (**a**) Case 1 of Fig. 3.5 (**b**) Case 2 of Fig. 3.5

performance of *Willow* is not significantly better than when the QoS aware scheduler alone is used. However at high utilizations *Willow* performs better than the case when the QoS aware scheduler alone is used since there are no adaptation related migrations. Figure 3.6b shows the percentage of queries with delay violations in Case 2. It can be seen that the benefits of Willow are very significant in Case 2 as compared to Case 1, especially at moderate to high utilization levels. Figure 3.6 shows that an efficient QoS aware scheduler alone cannot do any good in the presence of energy variations. Willow significantly improves the possibility of meeting the QoS requirements with the help of systematic migrations.

Figure 3.7 shows the percentage of queries with delay violations when the ambient temperature of servers 1–4 is 45 °C at 60% utilization. As explained before, at a moderately high utilization level (60%), Willow migrates applications away from high temperature servers and hence they run at lower utilizations. This in turn reduces the number of queries with delay violations.

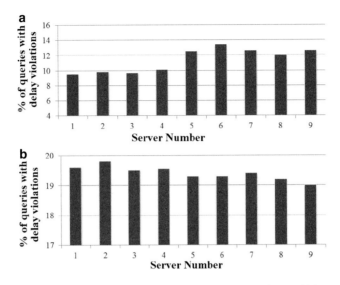

Fig. 3.7 Percentage of queries with delay violations when servers 1–4 are at high temperature at 60% utilization

3.3.2 Energy Adaptive Computing in Multi-tiered Datacenters

In this section we investigate a specific incarnation of *EAC* which is the manifestation of Energy Adaptive Computing in datacenters hosting multi-tiered web services [21]. These include a wide variety of applications ranging from e-commerce services to large scale web search applications.

The energy profiles of renewable energy resources provide ample opportunities for predicting the available energy during different time periods. For instance, the energy available from a solar panel may be correlated with the temperature (e.g., more solar energy during hotter days). The energy available from a wind power plant can be inferred from the weather forecast. Such interactions can be exploited in the adaptation mechanisms. If energy availability is predicted to be low during a certain period of time in a data center located at a specific location, some work can be done during the previous energy plenty periods. For example, in a datacenter supporting a web search application, background operations like web crawling and indexing operations can be done during energy plenty periods. Workload can also be migrated from datacenters located in places where there is surplus/cheap energy available. As simple as it may sound, the required control actions are complex and need to be continuously coordinated across multiple time granularities.

Figure 3.8 shows such a scenario with different time constants for an example datacenter with multiple clusters in different geographical locations. The time interval between successive control actions decreases as we move down the pyramid. In a datacenter which has hierarchical power distribution units (PDUs), the datacenter level PDU is at the top and the individual nodes are at the bottom of the hierarchy.

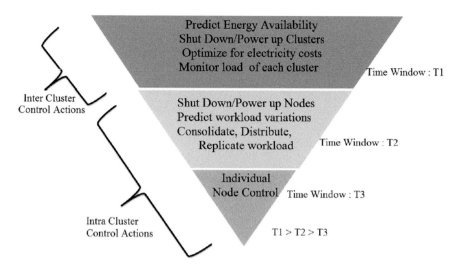

Fig. 3.8 Time windows for different power control actions

We classify the control actions into two, based on the direction of initiation of the control actions – those that are initiated from the top–down and those from the bottom–up along the hierarchy. The largest time window $T1$ in Fig. 3.8 represents the time interval during which the power supply varies. The variations are significant enough to initiate the control actions. The control actions may include predicting the available power for the next control period and migrating load from energy deficient clusters. In the multi-tier web server scenario that we consider, assuming that the services are stateless, workload migration involves redirecting more traffic to data centers with surplus energy. The number of servers in each cluster needs to be adjusted depending on the available power. Workload needs to be redistributed based on the number of servers that are kept powered on after the execution of the control actions. These are the top–down actions based on the available power. The bottom–up control actions are initiated from the node level. For instance some of the nodes might experience thermal constraints due to inefficient cooling or high thermal output. The demand variations and thermal constraints of the nodes need to be continuously monitored and reported to the managing entity (e.g., tier level load dispatcher) and the load and power budgets of the nodes need to be adjusted accordingly. This happens at a smaller time granularity ($T2 < T1$) than the power variations. At an even smaller time granularity ($T3 < T2$), the control actions may include adjusting the operational frequency of individual nodes or putting the nodes in shallow sleep modes if available.

The number of servers that are kept powered on and running in each tier is determined based on the available power and the delay constraints of the application. We model the problem of determining the servers that need to be kept powered on as a knapsack problem. The expected delay with a given number of tiers in

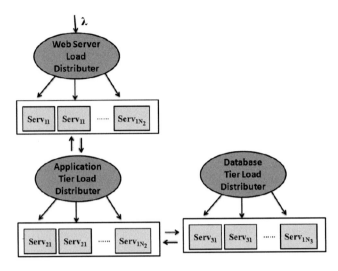

Fig. 3.9 Architectural assumptions of a three tiered datacenter

each tier is estimated based on queuing theoretic models and the best configuration that minimizes delay violations is chosen. Once this is done, some servers need to be turned on and some others need to be turned off. These operations are not instantaneous and involve significant overheads. For instance, before a server can be turned off, the pending queries that it was serving need to be completed and the workload that it was handling before needs to be redistributed to other servers. When a server that is currently powered off needs to be turned on, the activation period may be in the order of several seconds. An efficient planning of these operations is essential to avoid any adverse impacts on QoS. Many research works in the past have focused on formulating the power/performance trade-offs as optimization problems and have proposed solutions that would minimize these costs. However they fail to analyze and design an execution plan for the implementation of such optimization solutions. In order to design an execution plan, we leverage on a well known technique that is widely used for solving planning problems in Artificial Intelligence. A load dispatcher at each tier adjusts the load allocated to different servers in the tier. All these actions are executed at different time granularities and help in adapting to the available power profiles and the workload variations for efficient operation of data centers.

3.3.2.1 Architecture of a Multi-tiered Datacenter

Figure 3.9 shows the typical architecture of a data center hosting a web service application like an e-commerce service with three tiers. The frontend nodes in Tier 1 process the HTTP requests and forward queries to the nodes in application tier (Tier 2). The application servers handle the queries and send the appropriate

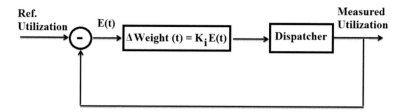

Fig. 3.10 Integral controller design for load dispatcher

response back after communicating with the database tier (Tier 3) which consists of a clustered database handled by multiple nodes.

Application Tiers

The front end nodes then process the response and format the response (e.g., convert to HTML) to send it back to the user. Each query thus passes through the same tier multiple times. We assume that the servers in each tier are stateless from the application point of view and any server can process any query. However, when a query from the same session revisits a tier, it needs to be serviced by the same server that handled it before. The load in the datacenter depends on the number of sessions in progress and the number of servers involved in processing the queries. QoS guarantees are typically expressed in terms of the overall response time which is the time elapsed between the user request and the time when the user gets back the response. Recently, newer datacenter architectures have been proposed where a large number of cheap commodity servers replace fewer powerful and expensive servers [22]. This is not only a power efficient architecture but also reduces the initial capital costs significantly. We consider such a scenario where there are multiple commodity servers in each tier and are typically homogeneous. Each server has a service queue where the queries are queued and a load dispatcher in each tier assigns the queries to the appropriate servers based on their capacity and current load.

The service rate μ_j^i (of server j in tier i), is different for different servers depending on power budgets and thermal constraints. For instance even with sufficient power budgets certain servers can only be operated at half their capacity due to inefficient cooling (e.g., servers in the top of racks) and the service time of a server running at 50% of its maximum frequency is almost double that of a server running at its maximum frequency of operation (ignoring memory access and other delays). The relationship between the power consumption and thermal limitations of a server was derived in our previous work [5] and is given by Eq. 3.2. Power budgets are allocated to servers so that the thermal constraints of individual servers are not violated and the capacity of servers are proportional to the power budgets.

Figure 3.10 shows the design of the load dispatcher. The load dispatcher in each tier accounts for the differences in capacities of the servers and adjusts the input to each server in the tier. The load dispatcher periodically infers the average service times of the individual servers and the current arrival rates for each server. The average utilization for tier i is then given by $\rho^i = \lambda^i / \mu^i$ where λ and μ are the average arrival rate and service times of the tier. The dispatcher in each tier i then adjusts the arrival rate of queries λ^i_j, for each server j so that the utilization $\rho^i_j = \lambda^i_j / \mu^i_j$ is the same ($= \rho^i$) for all servers j. The service rate reduction can be due to a number of reasons. The servers might be running at lower operational frequencies by means of techniques like *dynamic voltage and frequency scaling (DVFS)* [23]. An alternative technique is to force the servers to go to deep sleep states with low power consumption periodically in order to reduce their power consumption or prevent their temperatures from increasing beyond certain limits. Irrespective of the control mechanism used, the load dispatcher has to make sure that the load is shared proportional to the capacities (service rates) of the servers. During each integral control period ΔT_I the utilization of the tier is measured. The integral controller shown in Fig. 3.10 periodically adjusts the weights for each server depending on the difference between the target ($\rho_i = \lambda_i / \mu_i$) and measured utilizations. The input to the controller is the measured utilization of each server. The error term $E(t)$ is the difference between the target and measured utilizations at time t. The new weights for the time period $(t, t + 1)$ is given by the following equation.

$$Weight(t + 1) = Weight(t) + K_i E(t) \qquad (3.8)$$

where K_i is the integral constant. The arrival rates of each server for that period is then adjusted proportional to the weights assigned to each server.

Data Storage Tier

The database tier typically has very high service times since it has to fetch data from the backend storage systems and disk accesses are orders of magnitude slower than memory. Traditionally very powerful and expensive servers are used in the database tier which connect to backend hard disk arrays via high speed Storage Area Networks (SAN). All servers in the database tier can access the data across the SAN. This is the shared disk model where data is shared between all the servers and is shown in Fig. 3.11a.

However recently, there has been a rising popularity of scale out frameworks to process large volumes of data [22] and NoSQL databases [24]. A large number of commodity servers are beginning to replace a few powerful servers. The data is typically stored on the local disks of the servers as shown in Fig. 3.11b. If a node needs data from another node, the data has to be accessed through the network.

In the shared disk model, the storage tier can be modeled as a separate tier. However when the disks storing the data are powered down to stay within the allocated power budget, the data access involves spinning the disk back on and then accessing the data. Let the disks storing the data be labeled as $D_1, D_2, \ldots, D_N$.

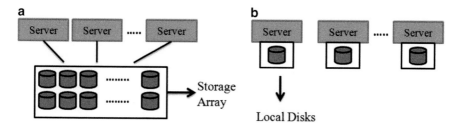

Fig. 3.11 Data storage models. (**a**) Shared disk storage model. (**b**) Shared nothing storage model

The disks that are already powered on are denoted by $D_1, D_2, \ldots, D_{N_1}$. Hence the average service time of the storage tier is given by Eq. 3.9.

$$\mu_{mean} = \sum_i \mu_{i1} P_i + \sum_j \mu_{j2} P_j, \quad i = 1 \, to \, N_1, \; j = N_1 + 1 \, to \, N \qquad (3.9)$$

where μ_{i1} is the service time of disk i that is already powered on and μ_{j2} is the sum of service time of the disk j that is powered down and the time taken to spin the disk back on from standby to ready state. Note that if all the disks are identical, $\mu_{i1} = \mu_1, \forall i$ and $\mu_{j2} = \mu_2, \forall j$. P_i is the probability of access of disk i. The number of active disks to be kept powered on is decided based on the power budget to the storage tier. However which disks to keep powered on depends on the popularity of data in the disks. For instance if there is a Zipf access distribution where only a subset of the data is the most popular, the disks storing the least popular data can be turned off so that the impact of spin up time delay is minimum. It is to be noted that this relies heavily on the data placement strategy. Previous works have proposed techniques that migrate the most frequently accessed data to a subset of the disks [25, 26]. These techniques can be complementary to our adaptation mechanisms.

In the shared nothing storage architecture model, when a node in the database tier is shut down, the data that it was handling is migrated to other nodes. The data migration involves a migration delay. Also there is a minimum capacity requirement for the stored data. Consequently, there are a minimum number of nodes that need to be powered on and hence there is a minimum power budget for the storage tier. Data migration across nodes involves large delays. Hence more care needs to be taken in planning the migration operations. We detail the planning of control actions in Sect. 3.3.2.3.

3.3.2.2   Delay Estimation (Mean Value Analysis)

An important step in determining the number of servers in each tier is to determine the overall delay across all tiers given the current workload demand of the data center. We leverage on queuing theoretic modeling of the datacenter to estimate the

Fig. 3.12 Closed queue
model of a datacenter

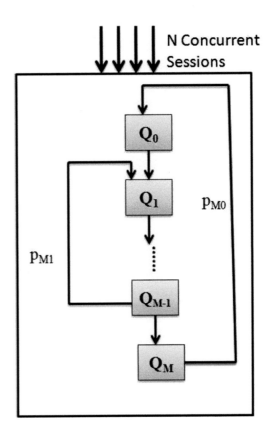

delays involved. Let us assume that the arrival rate of requests into the data center
is λ. As mentioned before the load dispatcher in each level i adjusts the arrival rate
to each server j in level i, λ_j^i so that the utilization $\rho = \lambda_j^i / \mu_j^i$ is the same for all the
servers. The requests pass through the same tier multiple times. The average number
of sessions in progress handled by the datacenter is N. With the above assumptions,
the multiple tiers can be analyzed as a closed network of queues [20, 27] as shown
in Fig. 3.12. The user requests originate from queue Q_0 and passes through each tier
multiple times. The delay at Q_0 corresponds to the user think time which is the time
spent by the user after receiving the response for a previous request and issuing
a successive request. Q_0 is an infinite server queue that generates N concurrent
sessions on the average. The traffic equation of each tier can hence be written as
follows.

$$\lambda^j = \sum_{k=0}^{k=M} \lambda^k p_{kj}, j = 1, 2, \ldots N \qquad (3.10)$$

where p_{kj} is the proportion of requests routed from queue k to queue j.

Mean Value Analysis (MVA) [20] is a popular technique for analyzing delays in networks of queues where each queue satisfies the $M \Rightarrow M$ property. This property states that if a queue is fed with Poisson input, the output process will also be Poisson. For a closed network, the MVA algorithm starts with the population of 1 and recursively computes the queue length and response time at higher population levels. The algorithm is shown in Algorithm 1.

Algorithm 1 MVA algorithm

$n_i(0) = 0, i = 1, 2, 3 \ldots M$

$\tau_i(N) = \dfrac{1}{\mu_i} + \dfrac{1}{\mu_i} \times n_i (N-1), i = 1, 2, 3 \ldots M$

$\eta(N) = \dfrac{N}{\tau_0 + \sum_{i=1}^{i=M} \tau_i(N)}$

$n_i(N) = \eta(N)\tau_i(N)$ (Little's Law)

where, $\eta(N)$ is the throughput of the system with N customers
$\tau_i(N)$ is the average delay of the ith server when there are N customers in the system and
$n_i(N)$ is the average number of customers in queue i when there are N customers in the system

---

In the case of FCFS scheduling discipline, $M \Rightarrow M$ property holds only for exponential arrival and service time distributions (M/M queues). Since exponential service time is not practical, we use a simple modification to extend the MVA for queuing networks where the $M \Rightarrow M$ does not hold as suggested in [20]. The Nth arriving customer will find $n_i(N-1)$ customers in service in tier i of which $U_i(N-1)$ will be busy already receiving service from the server in tier i. The average residual time of the customers is given by

$$\gamma_i = s_i \frac{(1 + CV_i^2)}{2} \tag{3.11}$$

where s_i is the service time of tier i and CV_i is the coefficient of variance of service times. The response time of the customer is therefore given by the following equation.

$$\tau_i(N) = \frac{1}{\mu_i}[1 + n_i(N-1)] + U_i(N-1)(\gamma_i - s_i) \tag{3.12}$$

Hence this value of $\tau_i(N)$ can be substituted in Algorithm 1 to get a more accurate estimate of the delay values for the FCFS service discipline with a general service time distribution. The mean service time of each tier depends on the thermal and power constraints of the individual servers and the arrival rates (assigned by the dispatcher). As before, the thermal constraints are ignored temporarily and it is assumed that each server in the tier can run at full capacity. We model the scenario of replication in each tier similar to [27]. Each server in the tier has a service queue and the load dispatcher is assumed to be a perfect load balancer. Hence for different configurations i.e., different number of servers in each tier, MVA can be used to calculate the mean overall delay across all tiers for the requests. The configuration with the minimum power consumption that can satisfy the QoS guarantees for the requests is then chosen. For the chosen configuration, the knapsack problem instance is solved at each tier.

3.3.2.3 Planning and Execution of Energy Adaptive Control Actions

As explained before, at each time step ΔT_S the algorithm to optimize for power and performance based on the predicted power supply is executed and the number of servers that need to be powered on is adjusted. New servers need to be powered on and some servers need to be turned off. In either case, workload needs to be reassigned and the sessions that are already in progress need to be completed before the servers handling those sessions are turned off. All these operations cannot be done instantaneously and each of them has associated costs.

Let the initial state of servers at time t be $X(t)$ and the goal state given by the knapsack algorithm is $Y(t + T)$. Here the term state refers to the set of servers that are powered on and turned off and their workloads. The interval between time t and $t + T$ of length T is the planning horizon. A transition from state $X(t)$ to state $Y(t + T)$ involves multiple steps. An exploration of each possible state transition has a time complexity factor that is exponential in the number of servers involved. Hence we employ search heuristics that are well known and widely used in Artificial Intelligence for planning. One such technique is A^* search algorithm [28] that employs a greedy best first search technique to find the least cost path from the initial state to the goal state. The cost function $f(x)$ from one state to another consists of an accumulated direct path cost value to the next state $g(x)$ and a heuristic function $h(x)$. The next state that is chosen is the one with the minimum cost function $f(x)$. The sufficient and necessary condition for optimality is that the heuristic function $h(x)$ is admissible. An admissible heuristic is one that does not overestimate the cost to the goal state.

Figure 3.13 shows the different state transitions during the execution of the required changes for adaptation. Let us now consider the steps involved during state transition. Let $\{S_i\}$ be the set of servers that are currently serving the requests. Let $\{A_i\}$ be the set of servers that need to be shut down and $\{B_i\}$ be the set of servers that need to be powered on. When a server has to be turned off, no new requests are forwarded to the server and the sessions that it is currently serving need to be completed. Hence the workload of the server to be shut down will have to be redistributed to other servers that are already up, thereby increasing the average delay in those servers. The path to reach the goal state from the start state needs to minimize the time taken for the control actions to be completed.

The cost function $g(x)$ which is the cost of getting to state x is therefore the estimated time to reach the state x from the start state. The heuristic function $h(x)$ is given by the following equation.

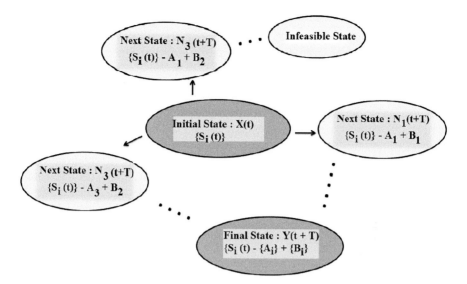

Fig. 3.13 State transitions from initial to goal states

$$h(x) = \max\left(\frac{Q_j(x)}{\mu_j(x)}, T_a\right), \ if \ N_A(x) \neq 0$$

$$= \frac{Q_j(x)}{\mu_j(x)}, \ if \ N_A(x) = 0 \qquad\qquad (3.13)$$

where j is the server with the maximum queue length in state x and $Q_j(x)$ is the queue length of j. $N_A(x)$ is the cardinality of the set of servers that are activated in state x. T_a is the time of activation of a server. $Q_j(x)/\mu_j(x)$ is the *virtual work* that needs to be completed before the server can be completely shut down. It can be easily seen that the heuristic function is never an overestimate of the actual delay involved since it is a conservative estimate of the most time consuming control action (deactivating/activating a server). Hence this heuristic function is admissible. Any state is considered infeasible if the power consumption during or at the end of the state is more than the power budget of the tier. A server with a long queue length, will most likely have a large workload to be redistributed and hence shutting it down will increase the utilization of the other servers. Hence it is better to mark it for deactivation towards the end of the planning horizon since most of the servers that are marked for activation would be fully functional by then and will be ready to accept the workload of the servers with long queues.

Note that the constraints that we introduce for the feasibility of a state is along the power dimension. Orthogonally, strict constraints can also be introduced along the time dimension and the power consumed can be minimized.

3.3.2.4 Simulation Results

As explained before we assume that the power supply consists of a constant source of power (henceforth referred to as *brown energy*/backup power) and a renewable (*green*) energy resource. Together, these two resources can support the datacenter operations at full load (≈ 50 kW). The reason behind this model is that in the current scenario, renewable energy resources alone cannot support datacenter operations entirely [29]. Hence we assume that when all the servers in the datacenter operate at full load, a portion of the necessary power to support the operation of the datacenter comes from conventional (backup) power resources. The available backup power is denoted as BP kW. The remaining power comes from a renewable energy source and is denoted as GP kW. Hence when the renewable energy source is operating at full capacity, $BP + GP \approx 50$ kW. In our simulations we assume that all of the available green energy is utilized first before using any power from the backup power resources so that the environmental impact of *brown* energy consumption is minimized.

We now investigate the energy profiles of renewable resources that our proposed scheme can successfully manage to adapt to. We changed the amount of the available backup power and estimated the extent of variations in renewable energy that can be tolerated i.e., the SLA requirement (delay $\leq 1,000$ ms) is not violated. The baseline scheme that we use to compare the efficiency of our scheme is one where the available power is allocated equally to all tiers. We choose this baseline scheme to demonstrate the effect of failing to account for the difference in delays in different tiers when allocating power budgets. With the baseline scheme, when there are 1,000 concurrent sessions, more than 100 servers need to be powered on to meet the SLA requirement. Figure 3.14a shows the level of variations that can be tolerated with our scheme. Let $TP = GP + BP$ be the total power required to support all servers in the data center. It can be seen that when the number of sessions is less (≤ 400), our scheme can successfully maintain the delay bounds even when the renewable energy varies by as high as 90% (i.e, the renewable energy generated is between 0.1 GP kW and GP kW) when the backup power is 20% of TP. Even when the number of sessions is as high as 1,000, the tolerable limit for variation in renewable energy is almost 47%. However in the case of the baseline scheme as shown in Fig. 3.14b, the tolerable variation limit is only 8% with a 20% backup power. This is because of the fact that the baseline scheme requires a lot more servers to be powered on than the MVA based allocation scheme in order to meet the delay bounds. It can be seen that as the proportion of backup power increases, the tolerable variation limit also increases as more power can be drawn from brown energy sources.

Figure 3.15 shows the proportion of green and brown energy required to support the datacenter operations when the available backup power is 20% of the total power required to support the entire datacenter ($0.2 \times 50 = 10$ kW in our case) and the renewable energy varies uniformly between 50 and 100% of the remaining 40 kW. We can see that when the number of sessions is small, the entire data center operations can be handled by renewable energy alone. When the number of sessions

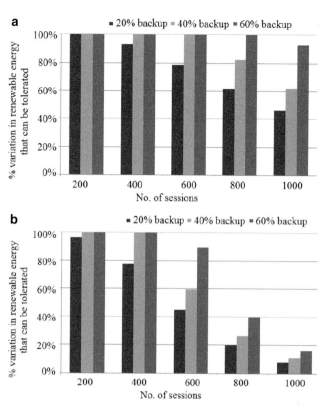

Fig. 3.14 Percentage of variation in renewable energy that can be tolerated for different proportions of constant power (**a**) Tier power allocation based on MVA (**b**) Equal power allocation to all tiers

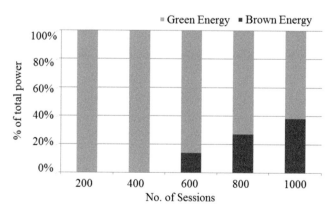

Fig. 3.15 Proportion of green and brown energy consumed when the renewable energy varies between 50 and 100%

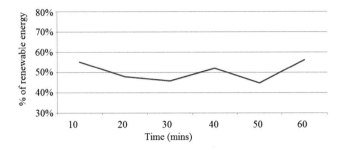

Fig. 3.16 Power supply variation assumed for simulation

increases, the variability in renewable energy requires that some backup power needs to be used occasionally. We can see that even when the number of sessions is as high as 1,000, only 38% of backup power is needed to guarantee the delay bounds. On the average, only 15% of the total power consumption comes from the brown energy source. This is because at each time instant in our scheme only the required number of servers are kept powered on in each tier to meet the SLA requirements. It is to be noted that we do not explicitly optimize the consumption of *brown* energy. Our scheme naturally reduces the energy consumption by powering on only the necessary number of servers in each tier to maintain the delay bounds.

Next, we evaluate the performance of our scheme under an energy constrained situation where the available energy is (at times) not sufficient to support the required number of servers to meet the SLA requirements. In order to simulate an energy constrained scenario, we assume the following power supply model. We consider a period of 1 h where the available green energy fluctuates between 0.45 GP and 0.55 GP kW which corresponds to a variation level of 55%. Note that if the variation in available green energy is $\leq 48\%$, then according to Fig. 3.14a there are no energy constraints and the delay bounds can be met with. Figure 3.16 shows the available renewable power profile that we assume for our simulation and the associated variations. It can be seen that the available renewable energy fluctuates every 10 min. With the presence of adequate energy storage infrastructure, these fluctuations may be at the granularity of hours.

Figure 3.17 shows the delay incurred when the number of sessions is 1,000 and the backup power is 20%. The power profile assumed is the same as shown in Fig. 3.16. The baseline scheme that we use here is one where the available power budget is equally shared among all tiers. It can be seen that when the power budget is allocated equally, the delay values are 48% more than our scheme on the average. A blind allocation of power to the tiers without accounting for the tier delays will result in high waiting times even when a large number of servers are powered on.

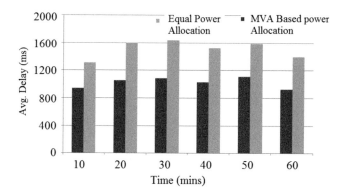

Fig. 3.17 Average delays with 20% backup power and 1,000 sessions

3.4   Conclusion

In this article we have discussed the challenges involved in adapting to the energy and thermal profiles in a data center. EAC puts power/thermal controls at the heart of distributed computing. We discussed how EAC can make ICT more sustainable and elaborated on two specific cases of EAC. First, we presented the design of *Willow*, a simple control scheme for energy and thermal adaptive computing in a datacenter running CPU intensive transactional applications. Second, we discussed the control actions that need to be executed at different time granularities and their implementation in a multi-tiered datacenter. We showed that an efficient implementation of the control actions can help reduce the associated overheads significantly. We have also demonstrated the ability of our schemes to adapt to significant variations in the available power supply. A major goal of this work is to inspire right-sizing the otherwise over designed infrastructure in terms of power and ensure the possibility of addressing the ensuing challenges via smarter control. A coordinated approach such as the one described in this article is necessary for coping with energy variability and variations in demand.

References

1. Heller B, Seetharaman S, Mahadevan P, Yiakoumis Y, Sharma P, Banerjee S, McKeown N (2010) ElasticTree: saving energy in data center networks. In: NSDI'10: Proceedings of the 7th USENIX symposium on networked systems design and implementation, San Jose
2. Gurumurthi S, Sivasubramaniam A, Kandemir M, Franke H (2003) DRPM: dynamic speed control for power management in server class disks. SIGARCH Comput Archit News 31(2):169–181
3. Chang J, Meza J, Ranganathan P, Bash C, Shah A (2010) Green server design: beyond operational energy to sustainability. In: Proceedings of the 2010 international conference on power aware computing and systems, ser. Vancouver Canada. HotPower'10

4. Kant K, Murugan M, Du D (2012) Enhancing data center sustainability through energy adaptive computing. ACM J Emer Technol Comput Syst 8:1–20
5. Kant K, Murugan M, Du DHC (2011) Willow: a control system for energy and thermal adaptive computing. In: Proceedings of 25th IEEE international parallel & distributed processing symposium, IPDPS'11, Anchorage
6. Greenberg S, Mills E, Tschudi B, Rumsey P, Myatt B (2006) Best practices for data centers: lessons learned from benchmarking 22 data centers. ACEEE summer study on energy efficiency in buildings, 2006, Pacific Grove
7. Flinn J, Satyanarayanan M(2004) Managing battery lifetime with energy-aware adaptation. ACM Trans Comput Syst 22:137–179
8. Kant K (2009) Challenges in distributed energy adaptive computing. In: Proceedings of ACM HotMetrics, Seattle
9. Krishnan B, Amur H, Gavrilovska A, Schwan K (2011) VM power metering: feasibility and challenges. SIGMETRICS Perform Eval Rev 38:56–60
10. Kusic D, Kandasamy N, Jiang G (2011) Combined power and performance management of virtualized computing environments serving session-based workloads. IEEE Trans Netw Serv Manage 8(3):245–258
11. Miettinen AP, Nurminen JK (2010) Energy efficiency of mobile clients in cloud computing. In: HotCloud'10. Boston
12. Chun B-G, Ihm S, Maniatis P, Naik M, Patti A (2011) Clonecloud: elastic execution between mobile device and cloud. In: Proceedings of the 6th conference on computer systems, ser. EuroSys'11. ACM, New York, pp 301–314. Available http://doi.acm.org/10.1145/1966445.1966473
13. Cuervo E, Balasubramanian A, Cho D-K, Wolman A, Saroiu S, Chandra R, Bahl P (2010) Maui: making smartphones last longer with code offload. In: Proceedings of the 8th international conference on mobile systems, applications, and services, ser. MobiSys'10. ACM, New York, pp 49–62. Available: http://doi.acm.org/10.1145/1814433.1814441
14. Petrucci V, Loques O, Mossé D (2009) A framework for dynamic adaptation of power-aware server clusters. In: Proceedings of the 2009 ACM symposium on applied computing, ser. SAC'09. Honolulu, Hawaii
15. Raj M, Kant K, Das S (2012) Energy adaptive mechanism for P2P file sharing protocols. In: CoreGRID/ERCIM workshop on grids, clouds and P2P computing, ser. EuroPar'12. Rhodes Island, Greece
16. Sharma A, Navda V, Ramjee R, Padmanabhan VN, Belding EM (2009) Cool-tether: energy efficient on-the-fly wifi hot-spots using mobile phones. In: Proceedings of the 5th international conference on emerging networking experiments and technologies, ser. CoNEXT'09. ACM, New York, pp 109–120
17. Friesen DK, Langston MA (1986) Variable sized bin packing. SIAM J Comput 15(1):222–230
18. VMWare vSphere, http://www.vmware.com/products/storage-vmotion/overview.html
19. Arlitt M, Jin T, 1998 World Cup Web Site Access Logs. http://www.acm.org/sigcomm/ITA/
20. Kant K (1992) Introduction to computer system performance evaluation. McGraw-Hill, New York
21. Murugan M, Kant K, Du D (2012) Energy adaptation for multi-tiered datacenter applications. Intel Technol J 16: 152–170
22. Barroso LA, Dean J, Hölzle U (2003) Web search for a planet: the Google cluster architecture. IEEE Micro 23:22–28
23. Dhiman G, Rosing TS (2007) Dynamic voltage frequency scaling for multi-tasking systems using online learning. In: ISLPED'07: proceedings of the 2007 international symposium on Low power electronics and design, Portland
24. NoSQL Databases, http://nosql-database.org/
25. Pinheiro E, Bianchini R (2004) Energy conservation techniques for disk array-based servers. In: Proceedings of the 18th annual international conference on supercomputing, ser. ICS'04. Malo, France ACM, New York

26. Colarelli D, Grunwald D (2002) Massive arrays of idle disks for storage archives. In: Supercomputing'02: proceedings of the 2002 ACM/IEEE conference on supercomputing. IEEE Computer Society Press, Los Alamitos, Baltimore, pp 1–11
27. Urgaonkar B, Pacifici G, Shenoy P, Spreitzer M, Tantawi A (2005) An analytical model for multi-tier internet services and its applications. In: Proceedings of the 2005 ACM SIGMETRICS international conference on measurement and modeling of computer systems, Banff
28. Russell S, Norvig P (2009) Artificial intelligence: a modern approach, 3rd edn. Prentice Hall, Englewood Cliffs
29. Stewart C, Shen K (2009) Some joules are more precious than others: managing renewable energy in the datacenter. In: Workshop on power aware computing and systems (HotPower) Big Sky, Montana

Chapter 4
Implementing the Data Center Energy Productivity Metric in a High-Performance Computing Data Center

Landon H. Sego, Andrés Márquez, Andrew Rawson, Tahir Cader, Kevin Fox, William I. Gustafson Jr., and Christopher J. Mundy

4.1 Introduction

4.1.1 In Pursuit of Energy Efficiency and Productivity

Rapidly rising energy consumption by data centers is an environmental and economic concern. The U.S. Environmental Protection Agency (EPA) conducted a comprehensive study of "energy use and energy costs of data centers and servers in the U.S.," including an appraisal of the "existing and emerging opportunities for improved energy efficiency" [8]. This study, in part, has influenced the efforts of the information technology (IT) industry to improve energy efficiency. For example, organizations targeting commercial data centers like The Green Grid (TGG) and the

This work is based on an earlier work: Implementing the Data Center Energy Productivity Metric, in *Journal on Emerging Technologies in Computing Systems*, (Volume 8, Issue 4, Article 30, October 2012) ©ACM 2012. http://doi.acm.org/10.1145/2367736.2367741

L.H. Sego (✉) • A. Márquez • K. Fox • W.I. Gustafson • C.J. Mundy
Pacific Northwest National Laboratory (PNNL), Richland, USA
e-mail: Landon.Sego@pnnl.gov; Andres.Marquez@pnnl.gov; Kevin.Fox@pnnl.gov; William.Gustafson@pnnl.gov; Chris.Mundy@pnnl.gov

A. Rawson
Advanced Micro Devices, Inc., Sunnyvale, USA
e-mail: Andy.Rawson@amd.com

T. Cader
Hewlett-Packard Company, Palo Alto, USA
e-mail: Tahir.Cader@hp.com

Uptime Institute are formalizing procedures and best practices to improve energy efficiency. Recent recommendations from ASHRAE's Technical Committee 9.9 [2], regarding acceptable increases in the ranges of operating temperatures, are designed to reduce cooling demands by exploiting thermal robustness of IT components. Major companies are promoting hardware and software technologies to address power and cooling demand in data centers. Besides the EPA, other government entities, including the Energy Efficiency and Renewable Energy office of the U.S. Department of Energy and national laboratories like Lawrence Berkeley National Laboratory and Pacific Northwest National Laboratory (PNNL) are developing tools and recommendations that are relevant to energy efficiency in the commercial and high-performance computing sectors.

The concept of *energy productivity* extends the notion of energy efficiency by considering the useful work accomplished by the data center in exchange for energy. Thus, if a data center produces the same amount of work using less energy than another, it has a higher energy productivity. Improving the energy efficiency, and perhaps more importantly, the energy productivity, of data centers requires the use of carefully chosen—and carefully applied—metrics. No metric is all-encompassing, and metrics can easily be miscalculated, misapplied, or misinterpreted.

4.1.2   The Challenges and Nuances of Measurement

A variety of approaches may be used to assess the performance of a data center in terms of energy efficiency, productivity, or both. Wang and Khan [34] provided a review of data-center metrics, a number of which we discuss here. One commonly used approach to assess efficiency is to compare the average power consumed by the entire data-center facility (including IT, cooling, and power delivery equipment) to the average power delivered to the IT equipment. Power usage effectiveness (PUE) and data center infrastructure efficiency (DCiE) are two such metrics [30]. Specifically, PUE is the ratio of total facility power to IT equipment power, and DCiE is the reciprocal. This practical approach, with some caveats, facilitates comparisons that are of interest to data center operators and designers.

However, these metrics have limited value to IT equipment users. Typically, users are interested in the performance of their applications, with the objective of minimizing the time to completion of a computing task or completing the maximum number of iterations of a program in an allotted period of time. Many metrics have been used to quantify these types of performance objectives. Examples include floating-point operations per second (FLOPS) and bits per second. These metrics tend to use artificial workloads to quantify completed computing work. Traditionally, these *component-level* metrics measure a capability subset of the machine that might not reflect the capabilities and shortcomings of the system as a whole.

As energy consumption has become an increasing concern, researchers have proposed metrics that compare the completed computational work to the energy required to carry out the computation. These include FLOPS per unit energy (FLOPS/E) and the energy-delay product (ED^α) [10, 12, 27]. Converting these metrics for use at the data-center level is problematic at best. If synthetic workloads are used, their performance is difficult to correlate to the performance of actual workloads at the data-center level because actual workloads tend to be a function of multiple inter-dependent performance aspects that are difficult to simulate. These aspects include processing performance, memory capacity and speed, network bandwidth and latency, and input/output (I/O) bandwidth and latency. Energy consumption by ancillary systems, such as transformers, uninterruptible power supplies (UPS), lighting, and cooling might not be properly considered. Consequently, it would be inappropriate to attempt to measure energy productivity at the data-center level using metrics like FLOPS/E or ED^α.

To overcome these shortcomings, new metrics are required at the data-center level that can bridge the gap between unit-less infrastructure efficiency metrics and component-level performance metrics. These metrics would address the needs of IT users interested in quantifying energy productivity from the perspective of their applications. Ideally, they would account for ancillary energy use and nonlinear computational effects.

Nonlinear computational effects on energy efficiency (e.g., inter-node communication, job scheduling, and load balancing, to name a few) pose a major hurdle in the definition of a useful metric that accounts for varying problem sizes. Similarly, nonlinear effects on energy efficiency at data-center scales—such as power distribution and cooling that are driven by ever-increasing power and heat density management requirements—must be accounted for in any useful metric applied at the data-center level.[1] Ad hoc solutions have extended component-level performance metrics to the data center with mixed success [10]. Users seem to agree vaguely on an understanding of the metrics (e.g., FLOPS/E), but comparisons across systems tend to be flawed and actionable measures to increase productivity are difficult to enact. Other attempts to develop data center productivity metrics will be explored in more detail in Sect. 4.2.2. Common to all these attempts is the realization that without a well-defined baseline (or a class of baselines), actionable productivity comparisons across systems is difficult at best.

4.1.3 Data Center Energy Productivity

We favor a user-centric approach to measuring the computational work performed by a data center. Instead of dictating performance rates to users (e.g., FLOPS, bytes/second), we propose that users define the performance units they deem important to make measurable progress towards job completion. Universality across

---

[1] Due to nonlinearity, we do not view productivity as an intensive property, as proposed by Kamil et al. [20]. However, we agree with their bias discussion: productivity can be biased by scale.

data centers and workloads is lost in this process—but for the data center manager, energy productivity improvements at the data-center level now become measurable, adapted to the particular workload of the data center. Towards this end, we decided to evaluate the DCeP metric, first introduced by TGG [29]. DCeP is the amount of useful work completed divided by the energy required to produce that work during a defined period of time.

The DCeP metric has a number of unique properties that make it realistic and useful in assessing energy efficiency at the data-center level. The physical measurements that constitute the numerator of DCeP (useful work completed) and the denominator (energy consumed) are gathered during an interval of time called the assessment window. Rather than simply measuring how much computational work is completed during the assessment window, the DCeP metric allows the user to define the computational tasks, transactions, or jobs that are of interest, and then assign a measure of importance or economic value to each specific unit of work completed. To the extent it is desirable and practical, the energy consumed during the assessment window may be measured at any point in the data center power distribution system. For instance, in addition to IT equipment, the measure of energy consumption may account for ancillary equipment such as power distribution panels, UPSs, pumps, motors, air handlers, and even the chiller plant.

Because the numerator and denominator of DCeP are obtained by integrating over the assessment window, it results in a more accurate assessment of energy consumption and productivity than would be obtained by one (or a few) instantaneous measures of performance rates and power consumption. Assessments based on a small number of instantaneous measurements may be prone to increased error and decreased precision[2] due to fluctuations[3] in these quantities over time. The assessment window for DCeP should be of sufficient length to mitigate the effect of rapid fluctuations—especially fluctuations in power consumption.

To evaluate the effectiveness of DCeP in quantifying the energy productivity of a data center, we used it to measure the outcomes of experiments conducted with the Energy Smart Data Center (ESDC) at PNNL. The ESDC was a highly instrumented data center constructed in 2007 for research in power-aware computing and data center productivity and energy efficiency. The ESDC's capability to measure chip, server, rack, and machine room performance holistically, as well as its capability to control cooling, was unusual in its expansiveness for a high-end computing data center. The fundamental objectives of these experiments were to demonstrate the practical use of DCeP in a data center, and determine whether DCeP could distinguish various operational states in a data center.

---

[2]Our use of the terms *accuracy*, *error*, and *precision* are consistent with those used by the International Vocabulary of Metrology [19].

[3]Figures 4.1–4.3 illustrate these fluctuations.

4.2   Data-Center Metrics

4.2.1   Considerations When Measuring Productivity and Efficiency

In general, defining the productivity and energy efficiency of a system is challenging due to confounding parameters that are open to interpretation and influenced by the context in which the system is used. These parameters include, among others, *system output/input*, *system scope*, and *system scale*.[4] The control-theoretic concept of system output/input might be characterized by several properties that are weighted according to interest (e.g., the output of computing performance given the inputs of energy and hardware). System scope is driven by responsibility, interest, and feasibility (e.g., whether energy use should account for IT equipment only, both IT and cooling equipment, IT and cooling and power distribution equipment, etc.). System scale classifies systems by size (e.g., office computing versus institutional computing). Programmable systems provide an additional twist by permitting the characterization of *workload* (the work mix that the system is required to handle). Perhaps most importantly, metrics should also be driven by *practicality* (the ease of implementation and interpretation of the metric). It would be ideal for productivity and efficiency metrics to account for each of these five parameters.

4.2.2   An Assessment of Various Metrics

We now briefly discuss a selection of productivity and efficiency metrics in terms of the parameters discussed in Sect. 4.2.1: system output/input, scope, scale, workload, and practicality. Debuting in November 2008, the Green500 [10, 32] re-ordered the top 500 highest-performing machines (Top500) according to a FLOPS/power metric, where IT equipment power was defined as input. Energy can be derived under the assumption that Top500 machines will be used to their fullest. By choosing to use an existing list of machines with an established workload of high-performance LINPACK (HPL), the authors emphasized practicality over all other qualities. Broadening the scope reduces practicality because measuring at the data-center or building level tends to be more onerous. Output, scale, and workload are predetermined by Top500 and, consequently, introduce intended and unintended biases (e.g., smaller HPL-optimized machines that still make it into the Top500 will generally come out ahead).

Addressing output and workload shortcomings that stem from the use of a single benchmark, the now disbanded SiCortex company introduced the Green Computing Performance Index (GCPI) in 2009 [34]. It broadened the Green500 workload with the High-performance Computing Challenge (HPCC) benchmark

---

[4]Similar concepts have been proposed by other organizations such as the Uptime Institute.

suite by intensively exercising memory and networks as opposed to focusing predominantly on the processors. The index was the sum of weighted contributions for each benchmark that were normalized against a reference system. The GCPI was less practical because it required that classes of suitable reference machines be identified for each system scale. Furthermore, estimating the system's HPCC power consumption while accounting for overhead and load factors is more involved than with HPL. While there does not appear to be any surviving documentation or references that describe the implementation of GCPI, we opted to mention it here to preserve the noteworthy idea of constructing a metric using multiple benchmarks.

The previously described metrics for high-end computing rank a list of machines according to productivity but do not highlight parameters that would assist in productivity improvement. Parametric models similar to Amdahl's and related laws [1, 15] help address these shortcomings by extending these laws with parameters that directly influence power consumption (e.g., load, frequency, and voltage) [11].

Anticipating power and energy productivity concerns before the high-end computing community, the commercial sector has been evaluating several metrics for some time. In 2005, Sun Microsystems unveiled SWaP [14, 34], a ratio of performance to space × watts. The performance is measured as system output (e.g., the number of transactions per second) and the space and watts are measured as system input. The declared intention is to reward the density of IT equipment. Scope and workload have to be specified to allow cross-system comparisons. In 2008, Standard Performance Evaluation Corporation (SPEC) introduced its first performance and power benchmark for workloads consisting of server-side Java [27]. SPEC provides a methodology to measure at different machine-load levels, thereby enhancing its practicality. Scope and scale are kept vague, but the tester is encouraged to document experimental provenance in detail.

Circumventing the difficulties in identifying and assessing relevant workloads, efficiency metrics measure overhead, losses, and energy use at multiple levels and scales of the data center. The most widely accepted efficiency metrics are PUE and DCiE, promoted by TGG [30], as well as the five metrics described by the Uptime Institute: site infrastructure energy-efficiency ratio (SI-EER), site infrastructure power-overhead multiplier (SI-POM), hardware power-overhead multiplier (H-POM), deployed hardware utilization ratio (DH-UR), and deployed hardware utilization efficiency (DH-UE) [28]. In the government sector, the U.S. EPA has specified ENERGY STAR server requirements [9].

4.2.3 Formal Definition of Data Center Energy Productivity

DCeP provides a measure of the useful work performed by a data center relative to the energy consumed by the data center to perform this work. Informally, we can express this as

$$\text{DCeP} = \frac{W}{E_{Total}} = \frac{\text{Useful work produced}}{\text{Total energy consumed by the data center}}. \tag{4.1}$$

DCeP should be computed over a contiguous time interval, called the assessment window. The assessment window should be defined to best suit the needs of the investigation at hand. It should be long enough to gather a representative sample of the workload of interest, including an adequate representation of the typical fluctuations in work and energy consumption. However, it should not be so long as to be impractical. For example, assessment windows may be selected to coincide with times of maximum computational task loading, or they may be chosen to focus on specific workload characteristics on certain days of the week.

Useful work is measured in terms of useful computational units (UCUs), and W is essentially a weighted, normalized count of the number of UCUs produced by one (or more) application(s) during the assessment window. Each UCU represents a discrete amount of work, such as an email transaction, the execution of a query, the completion of a simulation, etc. The UCU must be defined specifically for each task or application, and its relative value may be based on time and/or cost. We present a formulation of W that differs slightly from the notation used by TGG [29] to more clearly account for the work produced by multiple applications:

$$W = \sum_{j=1}^{N_a} \sum_{i=1}^{M_j} V_j U_j (t_{ij}, T_{ij}) C_{ij}, \tag{4.2}$$

where $j = 1, \ldots, N_a$ indexes the applications, N_a being the number of applications, and $i = 1, \ldots, M_j$ indexes the UCUs initiated by the application *during* an assessment window. M_j is the number of UCUs initiated during the assessment window by the jth application, V_j is the relative value of a UCU produced by the jth application, and $C_{ij} = 1$ if the ith UCU from the jth application completes during the assessment window and $C_{ij} = 0$ otherwise.

To account for the value of timely completion, $U_j (t_{ij}, T_{ij})$ is a time-based utility function for application j, where t_{ij} is the elapsed time from initiation to completion of the UCU and T_{ij} is the absolute time by when the UCU must be completed. We would typically expect the utility function to be decreasing in t_{ij}, and possibly going to 0 if the current time exceeds T_{ij}. Well-established techniques in decision analysis and utility theory [4, 7, 21] can be employed to determine the functional form for each U_j. There are also a variety of techniques available for determining the relative value weights, V_j [13, 22, 35, 36]. It is interesting that W closely resembles the additive utility function, a standard approach for calculating multiattribute utility [21].

Properly determining V_j and U_j for more than one application can be challenging because different applications may produce UCUs at different rates. Likewise, UCUs from different applications will likely represent different quantities of work. One application may have more intrinsic value than another, but assigning a specific value to each application may be difficult because "customers" of UCUs are likely to value output from some applications more than others. In these situations, the aforementioned decision analysis techniques for eliciting the V_j may be especially helpful.

Depending on the length of time required to produce UCUs, the assessment window should be long enough that the exclusion of UCUs that overlap the boundaries of the assessment window will not be a significant factor when making comparisons among different operational states in the data center.

In practice, the value of W is likely to be obtained by processing information in time-stamped log files that are produced by the applications under study. Our experience suggests that these log files typically require a fair amount of parsing and summarizing to arrive at the values of W for each assessment window, though the process could be automated easily. The details of calculating W for the experiment we performed are discussed in Sect. 4.3.5.

We obtained the value of E_{Total} by measuring the energy consumed by IT equipment and the corresponding cooling systems, including the energy consumed by the chiller plant and cooling towers that support the computer room air handlers (CRAHs) or air conditioners. This required frequent measurements of various systems in the ESDC and chiller plant. These measurements are described in more detail in Sect. 4.3.5 and the appendixes of Sego et al. [24].

4.2.4 DCeP and the Considerations of Productivity and Efficiency

We now consider how DCeP addresses the five parameters discussed in Sect. 4.2.1: system output/input, scope, scale, workload, and practicality. System output is a principal component of the DCeP metric, where the output, defined as useful work, is measured in terms of UCUs. System input for DCeP is the energy used by the data center, where the scope defines the extent of the energy calculation. Ideally, the largest possible system scope would be used in calculating E_{Total} to account for the energy consumed by IT equipment, the energy required to cool the data center, and losses associated with power delivery to both the IT and cooling equipment. However, the scope of the energy measurement may need to be restricted due to practical limitations in instrumentation at a particular facility. While the scope should be carefully defined and disclosed, the flexibility in defining the scope of the energy consumption does increase the practicality of the DCeP metric because users may find it difficult (or impossible) to measure all of the components at their facility that provide energy for data center operations.

While DCeP does not explicitly account for scale, the ratio of useful work completed to energy consumption allows for the comparison of systems[5] with different scales. Specifically, larger systems, which presumably consume more energy than smaller ones, must produce more useful work to have comparably high energy productivity. The versatility of the definition of W permits the assessment

---

[5]Provided the definition of useful work and the scope of energy consumption is consistent for all systems under comparison.

of virtually any system workload, provided the applications of interest can be measured in terms of UCUs. On the other hand, the subjectivity in defining UCUs and their associated relative values (V_j) and utility functions (U_j) will likely render DCeP less effective in making comparisons from one data center to the next unless standardized workloads and common definitions of V_j and U_j can be agreed on by interested parties. However, for the very reason of its flexibility, we believe DCeP may prove to be a practical measure for data-center operators who seek to improve within-data-center energy productivity by making within-data-center comparisons of various operational configurations within the data center. The practicality of DCeP may be improved by using alternative measures, or proxies [31], of useful work that may be easier to obtain than W as defined in (4.2).

4.3 Methodology

4.3.1 The Energy Smart Data Center

The ESDC was established at PNNL in 2007 with funding from the National Nuclear Security Administration of the U.S. Department of Energy. It consisted of a liquid-cooled (spray-cooled) IBM x3550, 9.58-TFlop cluster called NW-ICE that contained 192 servers, each with two 2.3-GHz Intel (quad-core) Clovertown CPUs, 16 GB DDR2 FBDIMM memory, 160 GB SATA local scratch, and DDR2 InfiniBand NIC. Five racks in the cluster were equipped with evaporative cooling at the processors, while two racks were completely air-cooled. The five liquid-cooled racks each contained a thermal management unit (TMU) that extracted heat from the liquid used to cool the processors and transferred that heat to a chilled water line. NW-ICE employed a Lustre global file system with 34 TB mounted and 49 TB provisioned. It was housed in PNNL's Environmental Molecular Sciences Laboratory (EMSL). The EMSL is a mixed-use facility that also houses the main cluster of the Molecular Sciences Computing Facility, providing an industry-relevant setting in which we could study data-center productivity and energy efficiency. The ESDC and its corresponding cooling and power equipment are discussed in greater detail by Sisk et al. [25]. The unique measurement capabilities of the ESDC exceeded those usually found at the single-component level or indirectly derived via performance profile estimation [10,11,27]. Comparable large-scale measurement harnesses include, for example, HP Data Center Smart Grid [16], HP Insight Control [17], and IBM Tivoli [18].

Using the ESDC, we conducted an experiment in November 2008 with the principal objective of determining whether DCeP could distinguish different operational states in the data center. These experiments were data-intensive and required extensive instrumentation, real-time access to sensor measurements, and semi-automated analysis routines to monitor system health (air and chip temperatures, humidity, power consumption, etc.), iteratively adjust experimental conditions, and rapidly query, summarize, and visualize data.

4.3.2   The High-Performance Computing Workload

As part of the experiment, an HPC workload consisting of two research applications
was processed by NW-ICE: the Weather Research and Forecasting Model (WRF)
and CP2K, an open-source program for performing atomistic and molecular simulations of solid-state, liquid, molecular, and biological systems.

WRF [26] is a state-of-the-art weather forecasting model used by many in-
stitutions around the world for production weather forecasts as well as detailed
process studies for basic research. WRF represents the atmosphere as a number
of variable states discretized over regular Cartesian grids. It is somewhat I/O-
intensive, repeatedly writing out three-dimensional representations to construct
a time series of the state of the atmosphere. The WRF simulation set-up for
ESDC consisted of a basic weather forecast for the period beginning June 26,
2008 12:00 UTC for a region covering most of North and Central America using
15-km grid spacing. Output was performed every three model-hours, with each
output consisting of a 2.3-GB netCDF file. To simulate heavier use of NW-ICE
by WRF, multiple simultaneous copies of the same simulation were performed.
These represent multiple realizations of an ensemble, which are often generated for
production weather forecasts. Each simulation used a total of 184 processes (182
for computation and two for I/O). For this experiment, the initial and boundary
conditions, as well as the physics parameterizations, remained identical in all
simulations to maintain similar load balances and simplify the set-up for the tests.

CP2K is an open-source, all-purpose molecular dynamics code written in
FORTRAN 95 [5]. CP2K has an efficient implementation of density functional
theory (DFT) that allows for the integration of nuclear coordinates under the
influence of a quantum mechanical interaction potential [33]. The strength of
the CP2K implementation of DFT is that it allows one to study phenomenon in the
condensed phase that can be experimentally verified. Calculations from CP2K have
far-reaching implications for understanding the basic chemical physics underlying
heterogeneous reactions at interfaces. Our system in this study comprised 216 water
molecules, which was one of the larger systems to that date for performing the
required statistical mechanical sampling for aqueous interfacial systems [3]. The
output consisted of synchronous 75-MB files representing coordinates and energies
at 0.5-fs time steps. These files contained a record of the useful computational
output, which we discuss further in Sect. 4.3.5.

4.3.3   Experimental Design

To design our experiment, we began by identifying a number of factors that were
likely to influence the value of the DCeP metric. Each combination of these factors
represents a possible operational state in the data center. Examples of these factors
include: the application load balance (the fraction of the nodes devoted to each
application), the balance between memory and CPU (whether to use all or half of

Table 4.1 Description of the randomized complete block design, the treatment assignments, and the start and end times of the assessment windows. All events in this table occurred in 2008

Block	Launch time	Treatment	Assessment window	
			Start	End
1	6-Nov 23:00	75%WRF-AllCore	6-Nov 23:30	7-Nov 00:55
1	7-Nov 01:15	25%WRF-HalfCore	7-Nov 01:45	7-Nov 03:10
1	7-Nov 03:30	75%WRF-HalfCore	7-Nov 04:00	7-Nov 05:25
1	7-Nov 05:45	25%WRF-AllCore	7-Nov 06:15	7-Nov 07:40
2	7-Nov 08:00	25%WRF-HalfCore	7-Nov 08:30	7-Nov 09:55
2	7-Nov 10:15	75%WRF-AllCore	7-Nov 10:45	7-Nov 12:10
2	7-Nov 12:30	25%WRF-AllCore	7-Nov 13:00	7-Nov 14:25
2	7-Nov 14:45	75%WRF-HalfCore	7-Nov 15:15	7-Nov 16:40
3	7-Nov 20:00	25%WRF-HalfCore	7-Nov 20:30	7-Nov 21:55
3	7-Nov 22:15	75%WRF-AllCore	7-Nov 22:45	8-Nov 00:10
3	8-Nov 00:30	25%WRF-AllCore	8-Nov 01:00	8-Nov 02:25
3	8-Nov 02:45	75%WRF-HalfCore	8-Nov 03:15	8-Nov 04:40
4	8-Nov 08:00	25%WRF-AllCore	8-Nov 08:30	8-Nov 09:55
4	8-Nov 10:15	75%WRF-HalfCore	8-Nov 10:45	8-Nov 12:10
4	8-Nov 12:30	25%WRF-HalfCore	8-Nov 13:00	8-Nov 14:25
4	8-Nov 14:45	75%WRF-AllCore	8-Nov 15:15	8-Nov 16:40

the CPU cores on a node), the scheduling algorithm (largest first, smallest first, first-in/first-out, random, etc.), cooling scheme (air- or liquid-cooled), processor architecture (RISC, CISC, AP, GPU), memory size, connectivity (Ethernet versus InfiniBand), and operating temperature.

For our experiment, we chose to vary the application load balance (between CP2K and WRF) and the memory/CPU balance (full- or half-core) because these two factors were relatively easy to implement and did not require costly or time-consuming hardware adjustments. The first factor, load balance, had two levels: 75% of NW-ICE running WRF (with 25% allocated to CP2K) versus 25% of NW-ICE running WRF (with 75% allocated to CP2K). The second factor, the percentage of cores utilized for computation, also had two levels: 50 and 100%. We refer to the four unique combinations of load balance and core utilization as *treatments*—and these treatments constitute a 2^2 factorial design [6]. We subsequently refer to these four treatments as 75%WRF-AllCore, 25%WRF-AllCore, 75%WRF-HalfCore, and 25%WRF-HalfCore. To replicate the experiment, each of these four treatments was exercised during four, 9-h blocks of time.

The structure of the experiment is illustrated in Table 4.1. The order of the treatments within each block was randomized, constituting a randomized complete block design [6]. All four treatments are present in each block. We suspected that time-varying energy consumption at the EMSL might affect the energy efficiency of the chiller plant that supplies cooling to the ESDC and the entire EMSL facility. Consequently, the blocks were chosen to account for the potentially different energy use at the EMSL facility during various periods of the day. Specifically, Block 1 occurred in the late evening and early morning of a weekday (Thursday night/Friday morning), Block 2 took place during a weekday (Friday), Block 3 was late evening and early morning of a weekend (Friday night/Saturday morning), and Block 4 took place on a weekend day (Saturday).

4.3.4   Experimental Protocol and Time Line

As shown in Table 4.1, each treatment period in a block lasted 2.25 h and consisted of a sequence of steps that were scheduled ahead of time. Each treatment period began with the simultaneous launch of the CP2K and WRF jobs, followed by a 30-min stabilization phase. This was followed by an 85-min assessment window (for which DCeP was calculated), followed by a 5-min ending buffer after which the CP2K and WRF jobs were terminated. The treatment period concluded with 15 min of cool-down prior to initiating the next treatment period. The 30-min stabilization phase was included to allow the heat and cooling loads to stabilize within the data center, and to give the CP2K and WRF jobs time to initialize and begin completing their UCUs with regularity.

The length of the treatment period was chosen to accommodate four treatments within a 9-h block of time while still providing assessment windows that were long enough to permit the completion of a sizable number of UCUs—and thus minimize the impact of uncounted UCUs that overlapped the boundaries of the assessment window. In Fig. 4.1, the entire process is illustrated for Block 3, during which the power use for each of the NW-ICE compute racks is graphed over time. Similar patterns of power use were observed during the other three blocks. The full- and half-core treatments are easily distinguished in Fig. 4.1.

During the course of the experiment, we monitored a number of parameters, such as chip temperatures (DIMM and CPUs), evaporative water temperatures, rack power, chiller and pump power, chiller tonnage, CRAH cooling loads, etc., to ensure regularity and consistency. Analysis of these data was made possible by a real-time software data collection tool known as FRED [25] and a custom querying and visualization package we created in R [23]. During the 25%WRF-AllCore treatment of Block 2, certain sensor readings in the chiller plant were interrupted for about 4 min, during which time they returned values of 0. To calculate the energy use of the chiller plant during this period, these clearly aberrant values were removed from the data and we assumed the true, but unknown, values of these measurements remained steady during the period of interruption. Apart from this, no other abnormalities were observed in the data.

4.3.5   Calculation of DCeP

Calculating the useful work produced by NW-ICE during the experiment required the identification of the UCUs for each application under test. For WRF, the

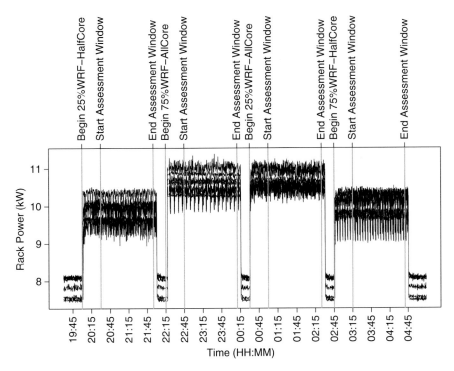

Fig. 4.1 Rack power consumption during Block 3. Each of the seven computing racks of the ESDC is represented by a single *black line*

completion of modeling 3 h of weather comprised the UCU, at which time WRF wrote an output file with an associated time stamp. The UCU for CP2K consisted of one full, self-consistent field optimization of the quantum mechanical wave function, resulting in the calculation of the energy nuclear forces for a particular arrangement of the simulated molecules. The completion times of these UCUs were consistently identified in the main log file of the CP2K program. Depending on the treatment, the number of UCUs that were initiated and completed during an assessment window ranged from 19–72 for WRF and 259–859 for CP2K. After accounting for the load balance and CPU utilization, the rate at which WRF produced UCUs was approximately 8% of the rate of UCUs produced by CP2K. Under the neutral assumption that the aggregate work accomplished by WRF during an assessment window was of equal value to the work accomplished by CP2K, we set $V_c = 0.08$ for all UCUs produced by CP2K and $V_w = 1$ for all UCUs produced by WRF, where the subscripts c and w are used to denote CP2K and WRF, respectively.

We also investigated the sensitivity of DCeP to various choices of V_c and V_w, which we discuss in Sect. 4.5. For simplicity, we set the utility function to be constant (i.e., $U_j(t_{ij}, T_{ij}) = 1$). Consequently, for our experiment, we can express the useful work from (4.2) as:

$$W = \sum_{j \in \{c,w\}} \left(V_j \sum_{i=1}^{M_j} C_{ij} \right). \tag{4.3}$$

The denominator of DCeP, E_{Total} (kWh), was calculated for each assessment window:

$$\begin{aligned} E_{Total} &= \frac{1}{0.97} \Big(\text{Energy consumed by ESDC} + \text{Energy required to cool ESDC} \Big) \\ &= \frac{1}{0.97} \Big(E_{NWICE} + E_{CRAH} \\ &\quad + \lambda_{Chiller}(H_{NWICE} + H_{CRAH}) + \lambda_{Tower} H_{TMU} \Big), \end{aligned} \tag{4.4}$$

where

- E_{NWICE} (kWh) is the energy consumed by the eight NW-ICE racks;
- E_{CRAH} (kWh) is the energy consumed by the two CRAHs in the ESDC;
- $\lambda_{Chiller}$ (kWh/Ton−hour) is the efficiency associated with the air-cooling in the ESDC (specifically, it is the energy efficiency of the entire EMSL chiller plant, calculated as the ratio of the aggregate energy utilized by the cooling towers, condenser pumps, heat recovery system, chillers, and chilled water pumps to the tonnage measured on the entire chilled water system);
- H_{NWICE} (Ton−hour) is the heat ejected to the air by NW-ICE that is removed by the CRAHs;
- H_{CRAH} (Ton−hour) is the heat produced by the motors of the two CRAHs (and removed by the CRAHs);
- λ_{Tower} (kWh/Ton−hour) is the predicted efficiency associated with the liquid cooling in NW-ICE, assuming (as discussed below) the warm water produced by the TMUs could be sent directly to the towers for cooling (thus, λ_{Tower} is the energy efficiency of the EMSL cooling tower system, calculated as the ratio of aggregate energy utilized by the cooling towers, condenser pumps, and heat recovery system to the tonnage measured on the water serving the chiller condensers);
- H_{TMU} (Ton−hour) is the heat extracted from the NW-ICE processors and ejected to water by the TMUs of the five liquid-cooled racks of NW-ICE; and,
- 0.97 is an estimate of the efficiency of the distribution of power from the utility company to the ESDC and the EMSL chiller plant and cooling towers.

Details regarding the calculation of each of these quantities are provided in the appendixes of Sego et al. [24]. While the ESDC and the EMSL chiller plant are highly instrumented, we were not able to measure the cooling capacity of the CRAHs directly (via water flow rates and changes in water temperature). Due to this and other constraints, we made the following assumptions to calculate E_{Total}:

1. After a period of stabilization, thermal balance existed between the heat ejected into the air by NW-ICE and the heat extracted by the CRAHs.
2. All electrical and mechanical power in the ESDC was dissipated as heat. This heat was ultimately ejected to the air in the ESDC and, for the liquid-cooled racks, heat from the CPUs was ejected to water.
3. Provided Assumptions 1 and 2 hold, the required cooling capacity for the CRAHs could be estimated reliably by the electrical power consumed by the CRAHs and by NW-ICE (after accounting for the heat evacuated by the TMUs directly to water).
4. The warm water returned by the TMUs can be discharged directly into the cooling tower condenser line without any pumping costs, and the cooling tower was capable of sufficiently cooling the water that is supplied to the TMUs.[6]
5. The room where the ESDC was housed had heating, ventilation, and air-conditioning (HVAC) vents and returns whose temperature and flow rates we were unable to control or reliably measure. Consequently, we ignored the cooling capacity of the HVAC system and assumed its cooling effect was negligible, or at least consistent, for each of the treatment runs that were conducted.

To support the first assumption of thermal balance, we adjusted the thermostat settings of the two air handlers to ensure that both did not run at 100 or 0% load simultaneously. These adjustments were made throughout the course of the experiment to accommodate the different treatment conditions, but not during the assessment windows. Ideally, both of the CRAHs should operate between 0 and 100% load, which was the case most of the time. To illustrate, the CRAH loads of the third block are displayed in Fig. 4.2, which were typical of the other three blocks. For each treatment period, the loads stabilized by the time the assessment window began. Initial thermostat settings for the two CRAHs were determined based on results from previous pilot experiments. Incidentally, determining the ideal thermostat settings for the two CRAHs proved to be quite challenging.

The energy efficiency of the entire chiller plant, $\lambda_{Chiller}$, and the cooling tower system, λ_{Tower}, varied during the course of the experiment. This is illustrated in Fig. 4.3. To ensure that changes in the chiller plant efficiency were not confounded with the effect of load balance and CPU utilization, we calculated these efficiencies over the course of the entire experiment, rather than calculating the efficiency for each assessment window separately. In so doing, we assumed these fluctuations in chiller plant efficiency were not attributable to the ESDC because the total heat produced by the IT equipment and the CRAHs was consistently 217 Ton–hours for each block, which represented only about 6.4% of the total cooling capacity produced by the EMSL chiller facility during those same blocks of time.

The effect of calculating $\lambda_{Chiller}$ and λ_{Tower} for the entire experiment versus calculating them separately for each assessment window is illustrated in Fig. 4.4. When constant values of $\lambda_{Chiller}$ and λ_{Tower} (estimated over the course of the whole

---

[6]This capability did not actually exist in the ESDC. Instead, the water temperature for liquid cooling was regulated via a separate heat exchanger meant to simulate the cooling that would be provided by cooling towers.

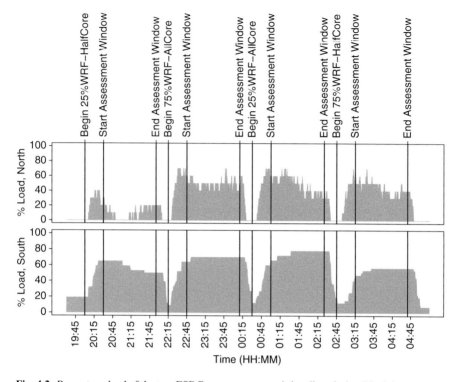

Fig. 4.2 Percentage load of the two ESDC computer room air handlers during Block 3

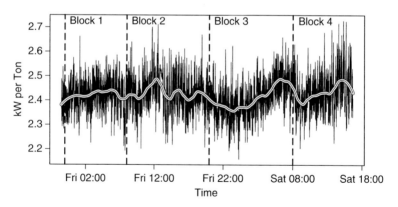

Fig. 4.3 Instantaneous total chiller plant efficiency during the course of the experiment, with *smoothing curve*

experiment) were used, the energy consumption was consistent from block to block, as illustrated in Fig. 4.4a. However, if $\lambda_{Chiller}$ and λ_{Tower} were calculated separately for each assessment window, we observed substantial changes in E_{Total} from one block to the next, as illustrated in Fig. 4.4b. While localized estimates

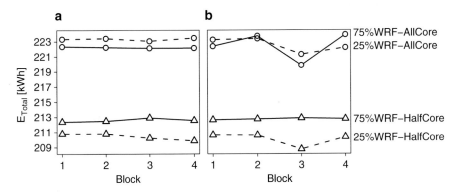

Fig. 4.4 Total energy consumption, E_{Total}, of the ESDC for each treatment and block. In panel (**a**), constant values of $\lambda_{Chiller}$ and λ_{Tower}, calculated from the entire experiment, were used to calculate E_{Total} for all assessment windows. In panel (**b**), $\lambda_{Chiller}$ and λ_{Tower} were calculated separately for each assessment window

of the efficiency (i.e., for only a particular assessment window) provided more accurate estimates of total energy use for that window of time, we wished to ensure that changes in these efficiencies did not influence our assessment of the performance of the various treatments. Consequently, in all subsequent analyses, we used $\lambda_{Chiller} = 2.42$ kWh/Ton−hour and $\lambda_{Tower} = 0.675$ kWh/Ton−hour, which were estimated using data from the entire experiment. The value of $\lambda_{Chiller}$ is considerably larger than λ_{Tower} because it includes the energy consumption of the entire chiller facility, including cooling towers, all relevant pumps, the chillers themselves, etc.

4.4 Results

The fundamental objective of the experiment was to demonstrate the practical use of DCeP and determine whether it could distinguish various operational states in a data center. To this end, we analyzed the effect of the treatments (load balance and CPU utilization) on W, E_{Total}, and DCeP.

Prior to calculating DCeP, we first counted the number of UCUs produced in each assessment window as defined at the beginning of Sect. 4.3.5. As expected, the UCU counts produced by a given treatment were nearly identical from one block to the next. The average number of UCUs for each treatment for both CP2K and WRF are given in Table 4.2. The range of the number of UCUs (the largest number of UCUs minus the smallest number of UCUs) observed for that treatment across the four blocks is shown to the right in parentheses. Naturally, the number of UCUs increased when the application was using a larger share of the data center and/or when a larger percentage of cores were used.

Table 4.2 Average and (range) of the number of UCUs, summarized over the four blocks

CP2K UCUs	Load balance		WRF UCUs	Load balance	
Percentage of cores (%)	75%WRF 25%CP2K	25%WRF 75%CP2K	Percentage of cores (%)	75%WRF 25%CP2K	25%WRF 75%CP2K
50	259.75 (1)	641.75 (2)	50	57.75 (1)	19.75 (1)
100	349.50 (3)	856.25 (6)	100	72.00 (0)	24.00 (0)

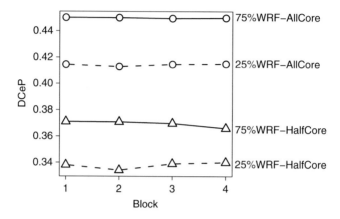

Fig. 4.5 DCeP for each treatment and block, with the relative value of a UCU from CP2K being 8% of the relative value of a UCU from WRF ($V_c = 0.08$ and $V_w = 1$)

Analysis of variance [6] of E_{Total} demonstrates a significant interaction (p-value <0.0001) between the load balance and the percentage of cores. Obviously, AllCore treatments consume more energy than HalfCore treatments. But the 25%WRF-AllCore configuration consumes more energy than 75%WRF-AllCore, whereas 25%WRF-HalfCore consumes less energy than 75%WRF-HalfCore.

Using the relative values discussed in Sect. 4.3.5 ($V_c = 0.08$ and $V_w = 1$), which presumed the work accomplished by CP2K and WRF during an assessment window to be of equivalent value, the DCeP metric showed clear distinctions among the treatments, as illustrated in Fig. 4.5. Analysis of variance showed the treatment effects to be highly statistically significant, with AllCore treatments having significantly higher energy productivity (DCeP) than HalfCore treatments (p-value <0.0001) and 75%WRF having higher DCeP values than 25%WRF treatments (p-value <0.0001). As with E_{Total}, DCeP remained constant over the four blocks for a given treatment.

Due to the consistency of E_{Total} and DCeP over time (i.e., across the blocks), we also considered the results by averaging across the blocks. This is illustrated by the interaction plots in Fig. 4.6. The interaction for E_{Total} is illustrated by the non-parallel lines in the plot, suggesting that the effect of the load balance of WRF versus CP2K depended on whether half or all the cores were used. However, for DCeP,

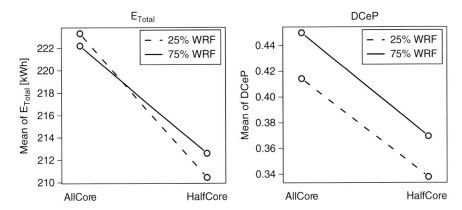

Fig. 4.6 Interaction plots showing the effects of load balance and core percentage on E_{Total} and DCeP, averaged over the four blocks. Each mean is significantly different from every other mean in that same plot

the interaction was not present (indicated by near-parallel lines), which led us to conclude that the effect of load balance on DCeP does not depend on the percentage of CPU cores. After averaging across the blocks, a Tukey pair-wise comparison test [6] demonstrated that each of the four E_{Total} treatment means for were significantly different from any other treatment mean (all adjusted p-values <0.005). The same holds true for the four DCeP treatment means (all adjusted p-values <0.0001). Of course, the separation in DCeP between the treatments is due in part to the choice of the relative value weights, V_c and V_w. We explore the sensitivity to these weights in Sect. 4.5.

4.5 Discussion

The sensitivity (or insensitivity) of DCeP to the choice of V_j and U_j should be understood by users. To illustrate, we examine the sensitivity of DCeP to the relative value weights, V_c and V_w, with six interaction plots shown in Fig. 4.7. In these plots, the relative value weight for CP2K was chosen to be 1, 5, 10, 20, 50, or 100% of the weight for WRF. The actual values of V_c and V_w were scaled to make the resulting DCeP values comparable while still achieving the desired ratio of V_c to V_w. Specifically, the weights were scaled so the mean DCeP score for the 75%WRF-AllCore treatment was always equal to 0.45 (the value obtained from the original analysis presented in Sect. 4.4 with $V_c = 0.08$ and $V_w = 1$).

While the relationship between AllCore and HalfCore treatments remained consistent regardless of the weighting scheme, the effect of the load balance between the applications was very sensitive to the choice of weights (a load balance of 25%WRF has lower DCeP values when CP2K UCUs were given less value relative

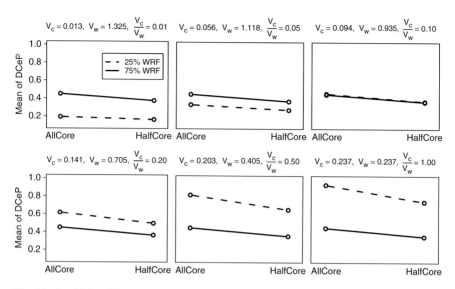

Fig. 4.7 Sensitivity of DCeP to the choice of relative value weights, V_c and V_w, where the weight for CP2K was 1, 5, 10, 20, 50, and 100% of the weight for WRF

to WRF UCUs), but the opposite occurs when CP2K UCUs were given more value relative to WRF UCUs. This sensitivity underscores the need to carefully develop a rational basis for choosing V_j when multiple applications are included in the DCeP calculation, especially when the applications produce UCUs at different rates with potentially different relative values, as was the case here. While we did not investigate a time-based utility function $U_j(t_{ij}, T_{ij})$, it stands to reason that W, and hence DCeP, will also be sensitive to the choice of the utility function.

Due to the sensitivity of DCeP to parameters that are chosen subjectively, its greatest value as a metric lies in comparing various operational configurations in the same data center. Using DCeP to compare different data centers would require a standard set of applications, the same choice of relative value weights and utility functions, and the same scope for energy calculations. In the absence of industry benchmarks, it would be difficult to compare the performance of two data centers using DCeP–especially if they belong to different organizations.

We found that implementing a designed experiment to study DCeP was advantageous for a number of reasons. The structure of the experiment with blocks over time led us to discover that the chiller plant and cooling tower efficiency were not as stable over time as we might have expected. This instability demonstrates the need for investigators to consider carefully the objective of their experiments when accounting for the energy required by chiller plants to supply cooling capacity to data center cooling equipment. That is, the fact that chiller plant efficiencies may change over time may obscure or confound the effects of experimental conditions that may be of principal interest, which was precisely the case for our experiment.

We addressed this issue by using a single estimate of chiller plant and cooling tower efficiencies that was computed using the entire period of experimentation.

Another benefit of using a designed experiment is that more than one factor can be investigated in a single experiment. This is more efficient than investigating single factors in separate experiments, especially because the multifactor approach allows researchers to identify potential interactions among the factors. For example, in our experiment, we investigated the effects of both application load balance and CPU utilization on DCeP and E_{Total}, which revealed an interesting interaction for E_{Total} (Fig. 4.6). Furthermore, randomization of the treatments helps avoid potential sources of bias from unknown factors that may influence the outcome of the experiment. Using a constant estimate for $\lambda_{Chiller}$ and λ_{Tower}, we found that W and E_{Total}, and hence DCeP, were very consistent (though not identical) over the four blocks. Consequently, for future experiments, we would likely use only two replicates (instead of four) and we may not use blocks. Nonetheless, the blocks proved valuable in demonstrating the consistency of the treatments over time and the instability in the chiller plant efficiency.

Experiments that are properly designed, executed, and analyzed can require considerable cost and effort. In particular, pilot experiments (which we conducted prior to the experiment described here) proved to be essential for identifying control settings and refining experimental protocols and logistics. Experimental costs must be weighed against the potential benefits of identifying optimal (or near-optimal) energy productivity configurations for data center operations.

4.6   Conclusion

We described the implementation of the Data Center Energy Productivity (DCeP) metric in an experiment conducted at Pacific Northwest National Laboratory using the Energy Smart Data Center (ESDC). We demonstrated that DCeP can be used to clearly measure and distinguish the energy productivity of different operational states in a data center. We investigated how DCeP is influenced by varying the load balance of two HPC applications and by simultaneously varying the percentage of cores (50 or 100%) utilized in the computations. Not surprisingly, we observed that 100% core utilization always resulted in higher energy productivity. However, the effect of the load balance on DCeP depended on the choice of the weights used to measure the relative value of the useful computational units (UCUs) for each application.

Even though the ESDC and its corresponding chiller plant were highly instrumented, we still found it necessary to make a number of simplifying assumptions to estimate the total energy consumption required to operate the ESDC. This challenge would be applicable to any metric for which one desires to measure energy consumption from the vantage point of the broadest possible scope (i.e., energy consumed by IT equipment and all ancillary equipment used to cool or otherwise support the data center). While it is a worthwhile endeavor to estimate the total energy consumption as accurately and with as broad a scope as possible, meaningful comparisons among operational states can still be made using DCeP with a narrower scope of energy use, provided the estimates of energy consumption are calculated consistently for each operational state.

Despite the subjectivity in the choice of relative value weights and the utility function that compose the numerator of DCeP, as well as challenges associated with calculating total energy consumption with a broad scope, the DCeP metric is useful for making comparisons of operational states within a given data center. These within-data-center comparisons can help operators identify hardware and/or software configurations that will improve energy productivity. The DCeP metric would have greater utility if it could be readily used to compare one data center to another. For any metric, defensible comparisons among data centers require a common standard of system output and input, workloads, and scope—while adjusting for scale. In addition to these parameters, using DCeP to compare data centers would require a common definition of useful work. Identifying approaches for comparing the energy productivity of data centers is worthy of future research and collaboration.

Acknowledgements This work was supported in part by the U.S. Department of Energy under DE-Award Numbers 47128, 55430, and SC0005365.

References

1. Amdahl GM (1967) Validity of the single processor approach to achieving large scale computing capabilities. In: Proceedings of the spring joint computer conference, AFIPS '67 (Spring), Atlantic City, 18–20 Apr 1967. Association for Computing Machinery, New York, pp 483–485
2. ASHRAE (2011) ASHRAE TC 9.9: 2011 thermal guidelines for data processing environments-expanded data center classes and usage guidance. Technical report, American Society of Heating, Refrigerating and Air- Conditioning Engineers. http://www.eni.com/green-data-center/it_IT/static/pdf/ASHRAE_1.pdf
3. Baer M, Mundy CJ, Chang TM, Tao FM, Dang LX (2010) Interpreting vibrational sum-frequency spectra of sulfur dioxide at the air/water interface: a comprehensive molecular dynamics study. J Phys Chem B 114(21):7245–7249
4. Berger JO (1985) Statistical decision theory and Bayesian analysis, 2nd edn. Springer, New York
5. CP2K (2011) CP2K developers home page. http://www.cp2k.org
6. Dean A, Voss D (1999) Design and analysis of experiments. Springer, New York
7. Edwards W, Miles R, von Winterfeldt D (2007) Advances in decision analysis: from foundations to applications. Cambridge University Press, Cambridge/New York
8. EPA (2007) Report to Congress on server and data center energy efficiency, public law 109-431. Technical report, United States Environmental Protection Agency. http://www.energystar.gov/index.cfm?c=prod_development.server_efficiency_study
9. EPA (2010) ENERGY STAR computer server specification Draft 1 Version 2.0. Technical report, United States Environmental Protection Agency. http://www.energystar.gov/ia/partners/prod_development/revisions/downloads/computer_servers/Draft1Version2ComputerServers.pdf

10. Feng W, Scogland T (2009) The Green500 list: year one. In: Proceedings of the 2009 IEEE international symposium on parallel & distributed processing, IPDPS '09, Rome. pp 1–7
11. Ge R, Feng X, Cameron KW (2009) Modeling and evaluating energy-performance efficiency of parallel processing on multicore based power aware systems. In: Proceedings of the 2009 IEEE international symposium on parallel & distributed processing, IPDPS '09, Rome. pp 1–8
12. Ge R, Feng X, Song S, Chang HC, Li D, Cameron K (2010) Powerpack: energy profiling and analysis of high-performance systems and applications. IEEE Trans Parallel Distrib Syst 21(5):658–671
13. Goicoechea A, Hansen DR, Duckstein L (1982) Multiobjective decision analysis with engineering and business applications. Wiley, New York
14. Greenhill D (2005) SWaP: space, watts, and power. Technical report, Sun Microsystems. www.energystar.gov/ia/products/downloads/Greenhill_Pres.pdf
15. Gustafson JL (1988) Reevaluating Amdahl's law. Commun ACM 31(5):532–533
16. Hewlett-Packard Company: HP Data Center Smart Grid. http://h17007.www1.hp.com/us/en/converged-infrastructure/ci-arch.aspx
17. Hewlett-Packard Company: HP Insight Control. http://h18013.www1.hp.com/products/servers/management/index.html
18. IBM: Tivoli Monitoring for Energy Management. http://www-01.ibm.com/software/tivoli/products/monitor-energy-management/
19. JCGM (2008) International vocabulary of metrology – basic and general concepts and associated terms (VIM). Joint Committee for Guides in Metrology. http://www.bipm.org/utils/common/documents/jcgm/JCGM_200_2008.pdf
20. Kamil S, Shalf J, Strohmaier E (2008) Power efficiency in high performance computing. In: IEEE international symposium on parallel and distributed processing, IPDPS '08, Miami, pp 1–8
21. Keeney RL, Raiffa H (1976) Decisions with multiple objectives: preferences and value tradeoffs. Wiley, New York
22. Ma J, Fan Z, Huang L (1999) A subjective and objective integrated approach to determine attribute weights. Eur J Oper Res 112:397–404
23. R Development Core Team (2011) R: a language and environment for statistical computing. http://www.r-project.org
24. Sego LH, Márquez A, Rawson A, Cader T, Fox K, Gustafson WI Jr, Mundy CJ (2012) Implementing the data center energy productivity metric. ACM J Emerg Technol Comput Syst 8(4):1–22 (Article 30)
25. Sisk DR, Khaleel MA, Márquez A, Hatley D, Cader T, Schmidt R (2009) Real-time data center energy efficiency at Pacific Northwest National Laboratory. ASHRAE Trans 115(Part I): 242–253
26. Skamarock WC, Klemp JB, Dudhia J, Gill DO, Barker DM, Duda MG, Huang XY, Wang W, Powers JG (2008) A description of the advanced research WRF Version 3. NCAR Technical Note NCAR/TN-475+STR, National Center for Atmospheric Research. http://www.mmm.ucar.edu/wrf/users/docs/arw_v3.pdf
27. Standard Performance Evaluation Corporation (2008) SPECpower_ssj2008 Benchmark. http://www.spec.org/power_ssj2008
28. Stanley JR, Brill KG, Koomey J (2007) Four metrics define data center "greenness". Technical report, The Uptime Institute.
29. TGG (2008) A framework for data center energy productivity. Technical report 13, The Green Grid. http://www.thegreengrid.org/en/Global/Content/white-papers/Framework-for-Data-Center-Energy-Productivity
30. TGG (2008) Green grid data center power efficiency metrics: PUE and DCIE. Technical report 6, The Green Grid. http://www.thegreengrid.org/en/Global/Content/white-papers/The-Green-Grid-Data-Center-Power-Efficiency-Metrics-PUE-and-DCiE
31. TGG (2009) Proxy proposals for measuring data center productivity. Technical report 17, The Green Grid. http://www.thegreengrid.org/en/Global/Content/white-papers/Proxy-Proposals-for-Measuring-Data-Center-Efficiency
32. The Green 500: http://www.green500.org

33. VandeVondele J, Krack M, Mohamed F, Parrinello M, Chassaing T, Hutter J (2005) Quickstep: fast and accurate density functional calculations using a mixed gaussian and plane waves approach. Comput Phys Commun 167(2):103–128
34. Wang L, Khan SU (2011) Review of performance metrics for green data centers: a taxonomy study. J Supercomput 1–18.
35. Wang YM, Luo Y (2010) Integration of correlations with standard deviations for determining attribute weights in multiple attribute decision making. Math Comput Model 51(1–2):1–12
36. Wang YM, Parkan C (2005) Multiple attribute decision making based on fuzzy preference information on alternatives: ranking and weighting. Fuzzy Sets Syst 153(3):331–346

Chapter 5
Sustainable Dynamic Application Hosting Across Geographically Distributed Data Centers

Zahra Abbasi, Madhurima Pore, Georgios Varsamopoulos, and Sandeep K.S. Gupta

5.1 Introduction

With the increasing prevalence of Internet-based computing services such as online gaming [7], cloud-based services [27], and search engines, the energy consumption in data centers to host such services has skyrocketed. Such increasing rate in energy consumption is of growing concerns to both operators and society. Electricity for Internet-scale systems costs millions of dollars, and burnt fossil fuels have detrimental impact on the environment. For these reasons and more, industry and research community propose to (i) increase data centers' overall energy efficiency [4, 10, 11, 18, 25, 28, 30, 38], and (ii) reduce data centers' dependence on fossil fuels [5, 29, 34, 45, 53]. Despite progresses, still the electricity cost is a huge concern to operators. According to a report by Intel Corp. and Microsoft [24], the energy cost accounts for over 10% of the total cost of ownership (TCO) of a data center.

This work has been partly funded by NSF, CRI grant #0855527, CNS grant #0834797, CNS grant #1218505 and Intel Corp.

This work is based on an earlier work: DAHM: A green and dynamic web application hosting manager across geographically distributed data centers, J. Emerg. Technol. Comput. Syst.(JETC) 8, 4, Article 34 (November 2012), 22 pages ACM

Z. Abbasi (✉) • M. Pore • G. Varsamopoulos • S.K.S. Gupta
Arizona State University, Tempe, AZ, USA
e-mail: zahra.abbasi@asu.edu; madhurima.pore@asu.edu; georgios.varsamopoulos@asu.edu; sandeep.gupta@asu.edu

P.P. Pande et al. (eds.), *Design Technologies for Green and Sustainable Computing Systems*, 117
DOI 10.1007/978-1-4614-4975-1_5, © Springer Science+Business Media New York 2013

Recently research community has proposed Geographical workLoad/appLication Placement (GLP) to shift workload toward data centers that offer lower electricity price or green energy at a given time.

Fortunately, *cloud computing* facilitates a dynamic, demand-driven allocation of computation and allows workload distribution across data centers. Applications can be assigned to *Virtual Machines (VMs)*, independent of the physical infrastructure. Virtualization provides a cloud the flexibility to host an application on the most cost-efficient data center at the time through VM migration.

5.1.1 Why GLP?

The ability to shift workload between data centers creates many energy management possibilities to lower electricity price, lower energy consumption, and efficiently manage the renewables [2, 3, 29, 31, 33, 40–43, 52].

"Follow the moon" takes advantage of lower costs for power and cooling during overnight hours. In this scenario, the workload is shifted across data centers depending on its local time to leverage low electricity cost from off-peak utility rates. In addition to the utility load, there are many other factors that cause electricity price to vary over time and location (e.g., hourly) including a variety of grid operators, power generation profiles, and wholesale markets. These factors all together contribute to the spatio-temporal variation of energy cost. These factors are shown in Fig. 5.1a.

Recently, energy buffering for minimizing energy cost in data centers has drawn attention [15,32,34,47,51]. The idea is to hoard energy in low utility rate periods (or when renewable is available) into batteries and draw from them during periods of high utility rate. GLP can jointly manage energy buffering and workload distribution to enable cost-efficient computation.

Beside energy cost reduction, importance of using green energy is increasing in data centers. Data center operators have started deploying renewables to partially power their systems. However, due to the unpredictability and fluctuation in availability of renewables, sustainability can hardly be achieved without large-scale batteries. GLP manages the computational load over a cloud by distributing it according to the availability of green energy at the time. The idea is to move computation across data centers, so as to minimize energy buffering, effectively reducing the overhead of energy storage.

In addition to the aforementioned factors, the heterogeneity of data centers in terms of computation speed, computing and cooling energy efficiency is another strong motivation toward GLP. In most data centers, computing servers are partially upgraded every 2–3 years. For example, in a 5-year old data center, several generations of equipment co-exist, thus resulting in heterogeneity, where servers have different computing capacity, power rating, and consequently, different computing energy efficiency. In addition to the magnitude of power, the power-utilization curve, specifying the power proportionality of servers vary among servers [6, 49]. An ideal power-proportional server consumes no power when idle as well as its power increases linearly with its utilization. However, current data center servers are not ideally power-proportional. Though the idle power of modern servers has reduced, the power utilization curve is not linear. Further, the energy efficiency, utilization over power consumption, of modern computing systems are becoming more diverse with respect to power characteristics [49].

Power Usage Effectiveness (PUE), which measures the efficiency of a cooling system and any source of power consumption other than the computing equipment in data centers, may vary among data centers as well. A large PUE is a strong indication of large cooling power, since the cooling system is the biggest consumer of the non-computing power in a data center (followed by power conversion and other losses). According to the US Department of Energy [14], a modern data center's PUE is around 1.7, which means that $0.7/1.7 \simeq 41\%$ of the power is, in its most part, consumed in cooling the data center. The data centers' thermal conditions and the type of cooler affect the cooling energy. Thermal condition of the data center room depends on the design of the room. The types of coolers that data centers use may depend on their location. Chillers are widely used as cooling systems in data centers which refrigerate water to cool the room but require a large amount of electricity to operate. To save power, many data centers are reducing their reliance on chillers and use the outside air to support the cooling systems [34]. Variety in data centers' room design and locations result into variety of PUEs.

All the above mentioned spatio-temporal variables make one data center the most cost-efficient at one time, and another data center at another time (refer to Fig. 5.1).

5.1.2   Applications' Requirements

For dynamically shifting workload across data centers, data center management should be aware of the network delay and bandwidth overhead during migration (e.g., user state data). This overhead depends on the type of applications, which can be either stateless or stateful. In *stateless applications*, e.g., search engines, the state of online users is not recorded; whereas *stateful applications*, e.g., multi-player online games, keep track of the state of users [7]. Therefore, stateful applications tend to induce higher migration cost.

Finally, *for Web applications, requests may originate from different locations or geographical areas*. As such, the network delay from these locations to the hosting data centers might also impact the end-to-end delay experienced by the users. Further, the bandwidth cost is different for different providers [41]. These may prevent an application from being hosted at certain locations.

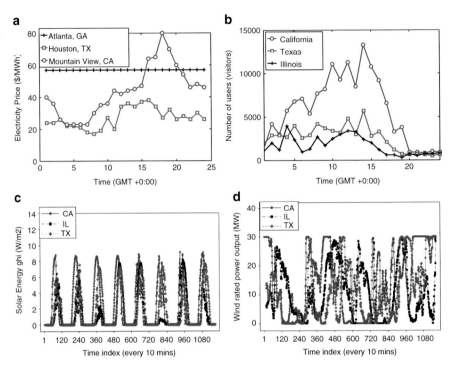

Fig. 5.1 Spatio-temporal variation of energy source and demands of data centers. (**a**) Hourly electricity price data for three major locations of Google IDCs on May 2nd, 2009 (data are taken from [43]), (**b**) hourly number of online users from three states for an entertainment Web site hosted at GoDaddy.com [3], (**c**) solar energy traces for three different sites, and (**d**) wind energy traces for three different sites

5.1.3 Challenges

Dynamic workload management across data centers is based on mathematical modeling and optimization algorithms to minimize energy cost under performance and the available (green) energy constraints. However, in practice it is quite often impossible to exactly model all the involved parameters (e.g., workload migration and battery characteristics) or know all of the required data (e.g., electricity price, workload demand) in advance. Further, the complexity of models such as performance and power make the problem non-convex and non-linear which suggest to approximately optimize energy by making simplistic assumptions about the models.

Besides technical challenges, there are some implementation challenges, including GLP management overhead and dynamic request direction that the solution should account for them. Sometimes data centers may belong to different companies. In this case, more management parameters such as accounting come into play.

The aforementioned challenges are partially resolved in recent literature. However, most of the efforts focus on improving/developing GLP problem under a specific workload type or data center energy management configuration. To our knowledge, there is no work that surveys GLP as a general problem to manage the heterogeneous workload over a cloud with heterogeneous energy infrastructure.

This chapter gives an overview on GLP. Further, it discusses the existing challenges and proposed solutions along with some important GLP modeling issues. GLP generally can account for any mix of interactive and batch jobs. However for the sake of simplicity, the given model assumes delay-sensitive stateless and stateful applications. The modeling of GLP is based on a cost optimization problem to minimize electricity cost and maximize the utilization of the available renewable energy profile under the constraints of applications' delay requirements, energy storage sizes, and servers' capacities.

Although GLP can potentially be a significant aid in handling the intermittency of renewable sources and electricity price fluctuations, a study that highlights the impact of GLP on the size of renewable infrastructure and batteries is still lacking. We perform a simulation study to evaluate the efficiency of GLP to make sustainable data centers. The study is performed using realistic traces and various data center configurations, e.g., workload and renewable energy predictability, battery sizes, and renewable energy profiles in data centers. Finally the chapter concludes and highlights future work and research directions.

5.2 Preliminaries

5.2.1 Wholesale Electricity Market

Electricity price vary over time and location. The variation is due to several factors including power generation, and more importantly supply-demand variation, and the market.

Electricity is produced from a variety of sources including coal, natural gas, nuclear power, and hydroelectric generation. Different regions use different sources depending on the availability of sources and their expenditure. For example in US. the total generation output in 2011 shows that coal dominates (40%), followed by natural gas (25%), nuclear (19%), hydro (8%), and renewables (5%) generation [14].

The key limitation of the electricity comes from the fact that it currently cannot be stored in a scalable and cost-efficient way. A sophisticated control is needed to ensure a close match between the supply and the demand. Any mismatch between the two can induce a high cost as power producers may need to add or remove the generation plants or load both of which are costly. To reduce such problems, system operators, known as balance authority, closely monitor the system to ensure capacity reliability. The system operators consisting of utilities, federal agencies and Independent System Operators (ISO) or Regional Transmission Organization

(RTO)s, forecast demand in day-ahead market, schedule power generation, reserve and transmission, adjust schedule as hours get closer, correct imbalances in real time, restore systems if disturbance occur and sometimes plan for long-term capacity and transmission upgrade.

System operators are usually regulated (e.g., by local government) to set rates, prescribe accountings, enforce reliability/safety and evaluate the need for new projects.

System operators in many regions of north America are ISO/RTOs which manage the grid. RTOs also administer wholesale electricity markets. The pricing in the wholesale market can be day-ahead, hourly basis or real-time.

The system price in the day-ahead market is determined by auctioning mechanism for the producers and the customers at each node to develop a classic supply and demand equilibrium price, usually on an hourly interval, and is calculated separately for subregions in the grid.

RTOs set the Locational Marginal Price (LMP) for different nodes in the grid which consists of three components: (i) System Energy Price (SEP): system clearing price if no congestion exists (always same at all locations), (ii) Marginal Lost Cost (MLC): Cost of marginal losses along transmission into specific node, and (iii) Marginal Congestion Cost (MCC): If congestion is positive, cost is incurred by expensive energy delivered to the destination. Whereas negative congestion indicates that the electricity generated is more than its demand. The cost is then calculated for each less MW that destination nodes consume compared to what is generated at source nodes in the grid.

The above discussion highlights the various parameters that affect the spatio-temporal variation of the electricity price which is a motivating factor for GLP. We also assume a time varying electricity price in this chapter.

5.2.2 Renewable Energy in Data Centers

Renewable energy are usually very expensive to implement, depend on the surrounding weather conditions, intermittently available, and require a big land area to implement in many cases. Despite drawbacks, data centers have already started to deploying them in various ways, not only to make their commitments for sustainability, but also to mitigate any steep raise in the electricity price in future. Google, Apple, FaceBook and many other industry leaders already made investments to partially or totally power their data centers from renewable energy sources [20, 21, 37, 50].

Some modern data centers have already installed (or are installing) on-site solar generation (e.g., i/o data center, FaceBook and Apple). This is in-spite of its high cost and the need for large arrays of Photovoltaics (PVs), to generate a small fraction of energy. Further, there are few examples of data centers that have installed wind turbines to power data centers from wind energy. Also some data centers, specifically small data centers, are seeking 100% on-site wind energy for their data center, e.g., Microsoft Virtual Earth. Furthermore, there are an increasing number of data center providers that use utility power that is sourced from wind generation e.g., Google.

Due to limitations of on-site renewable energy sources, i.e., geographical location or land, many companies do not have opportunities to install on-site renewable sources or directly use renewable utility power. There are other solutions such as Renewable Energy Certificates (RECs) that data centers can purchase to contribute in the growth of renewable energy industry. In this way, data centers support renewable energy producers by committing to buying their energy for long-term, but use brown energy in site [21].

In this study we only consider renewable energy sources that directly power data centers' servers. The source is assumed to be either generated on-site or purchased from a utility.

5.2.3 Sustainability Using Renewable and GLP

GLP manages workload and energy buffering across data centers rather than within a single data center. The result is to (i) shift the peak demand away from high electricity rate periods and push it into data centers that offer low electricity price or green energy at a time, (ii) store energy when the electricity rate is low or when the excess renewable energy is available and use it at other times.

A pictorial representation of GLP is presented in Fig. 5.2. GLP takes information of the workload, the available energy sources, and the battery state into account while placing the newly arriving workload. It also decides on the charging or discharging of batteries at each time instant considering the electricity cost and available renewable profile over time. However, in reality, various prohibiting factors such as battery's physical characteristics and workload migration overhead, may prevent GLP to utilize the low-cost energy. Frequent charging and discharging of a battery reduces its life time [16]. Further, the charging and discharging rate of a battery depends on its physical characteristics. Furthermore, there are many other factors such as energy density, power density, ramping time, energy efficiency and self-discharge that need to be taken into account to achieve a cost efficient solution [51]. Workload shifting incurs bandwidth cost and delay. GLP should avoid shifting workload from one data center to another, if its migration cost outweighs its energy saving due to shifting.

In this study, we assume ideal batteries for the sake of simplicity (we ignore the aforementioned physical limitations of batteries). The modeling in the next section accounts for workload migration overhead.

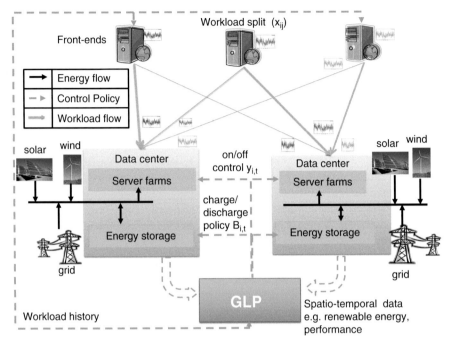

Fig. 5.2 Pictorial representation of GLP to achieve sustainable data centers

5.2.4 Practical Issues

GLP is based on the assumption that the underlying infrastructure allows request redirection mechanism. Request redirection are already in use to enable replication over Internet and Content Delivery Network [12, 39]. In this regard DNS based Request-Routing techniques are common due to the ubiquity of the DNS system. In DNS based Request-Routing techniques, a specialized DNS server is inserted in the DNS resolution process. The server is capable of returning a different set of records based on user defined policies, metrics, or a combination of both [9]. There are also other techniques such as HTTP redirection using persistent HTTP proxies to tunnel requests, which are currently employed for selecting data centers. These techniques usually incur high networking overhead (i.e., delay, bandwidth).

Although the concept of dynamic request redirecting is not new, more efficient techniques are required in order to ensure that GLP maintains the delay requirements of delay-sensitive application over a cloud.

Another practical issue is the ownership of data centers in a cloud. A cloud may consist of several data centers which are owned independently. Depending on the cloud accounting policies, another cost factor, in addition to the power cost, may affect the total cost of GLP.

For simplicity, GLP modeling in this chapter is based on the assumption that the cloud infrastructure supports request redirection and there is no accounting policies across data centers.

5.3   GLP System Model and Formal Definition

This section gives a system model and problem statement of GLP for interactive stateless/stateful applications. The model is inspired by the models and assumptions that are made in recent studies [2,3,29,31,33,40–43,52]. The goal is to show a cost model of GLP and discuss the associated challenges and solutions. The discussion, however, is not limited to the model of this section, but to a more general GLP problem which involves more practical phenomenon (e.g., mix of interactive and batch applications).

GLP for interactive jobs can be generally modeled as a network flow optimization model on a bipartite graph (see Fig. 5.2). End users' requests arrive from $|A|$ geographically distributed front-ends (i.e., the sources) where $A = \{a_1 \ldots a_j \ldots a_{|A|}\}$ denotes the set of front-ends (we use the term area and front-end interchangeably in the rest of this chapter). The geographical front-ends may be network prefixes, or even geographic groupings (states and cities). The reason to include multiple sources of workload (i.e., front-ends) is twofold. First, the bandwidth cost is an important contributor to data centers' TCO, and there may be large differences between costs on different networks, and sometimes on the same network over time [41]. Second, network delay between different front-end and data centers may vary over time and network. This may prevent some workload to be shifted to some data centers due to their delay requirements.

The workload must be distributed among the $|S|$ available data centers in the cloud (i.e., sink), where $S_t = \{s_{i,t}\}, i = 1, \ldots, |S|$ denotes the set of available data centers in the cloud, and each $s_{i,t}$ represents the number of available servers in data center i at time t. Also data centers may be provided with an energy storage of limited size, B^{size}, to smoothen the fluctuation in the availability of the renewables as much as possible.

There are many possible energy optimizations that can be developed by taking into account factors such as the workload split between the data centers, power state of the servers, migration overhead of the user state data, energy buffering levels and performance requirements of applications. For simplicity, we only focus on the workload split, a two power state for servers (active and off) as well as renewable energy buffering. The goal is to perform workload consolidation over minimal number of servers in the most cost-efficient data center at any given time. Extra servers are assumed to be turned off. Mathematical modeling of GLP consists of energy consumption model and cost, performance, workload and migration model of applications. The following sections introduce these models.

Table 5.1 Symbols
and definitions

Symbol	Definition
t	Epoch index
i	Index of data centers
j	Index of areas
$x_{i,j,t}$	Workload share of area j to DC i
$y_{i,t}$	Number of active servers
τ	Length of epochs (in second)
p_i^{idle}	Server idle power
p^{util}	Server peak power minus p^{idle}
c_i	Avg. util. of a user on a server
u_i^{th}	Threshold util. of servers
n_i^{th}	Affordable no. of users for servers
s_i	Number of available servers
d	Total delay of a request
d'	Data center delay
d''	Network delay between areas and DCs
d^{ref}	Total reference delay
d'^{ref}	Service reference delay
e_i	Electricity cost
β	Migration cost per migration
α	Switching cost of a new server
η	Performance violation cost per each user
$n_{j,t}$	Avg. number of online users in area j
si	Percentage of new users over an epoch
so	Percentage of users to sign out over an epoch

5.3.1 Performance Modeling

Performance constraints will enforce workload management to consolidate workload without compromising performance requirements of applications. Internet
applications are usually delay-sensitive such that their Quality of Service (QoS)
mainly depends on the end-to-end delay. For that we assume that in order for the
end users to experience a high QoS, their delay should not go above a *reference
delay*, d^{ref}. The delay d experienced by a user consists of the service delay d', i.e.,
data center delay, and the network delay d'', i.e., the delay between the front-end
and the data center; thus, $d = d' + d''$ (Table 5.1).

In Internet data centers, the SLA statistically bounds the delay, e.g.,, the delay
of q percent of requests should not go beyond the reference service delay d^{ref}. We
use the performance model by Chase et al. to guarantee the SLA for data centers
[1, 10]. This model asserts that the service delay strongly depends on the servers'
utilization levels; specifically, the SLA is guaranteed under a utilization threshold.
This threshold depends on the capacity of the servers and the type of the application.
Specifically, there is a threshold u_i^{th} associated with each data center's servers, such
that, if servers are not utilized above that point, the service delay d_i' respects the
SLA, i.e., $d_i' \le d'^{\mathrm{ref}}$ for at least q percent of users.

The above utilization-to-delay model can be replaced by queuing models, e.g., M/M/n or GI/G/n, mainly because the average delay is linearly correlated to the arrival rate λ and the per-request utilization ($d' = f \frac{\lambda}{d'=f(u)}$, where $u = \frac{\lambda}{service\_rate}$), based on Little's Law. In GI/G/n, the coefficients of variation of the workload arrival rate and of the service time come into play but they do not change the nature of the problem.

In modeling the network delay, we consider the delay differs depending on the network distance between the areas and the data centers. Also this delay may vary over time, depending on the network congestion. If we denote a delay as $d_{i,j,t}$ to mark the dependence on data center i, the front-end j and the epoch t, it will be sum of the network delay and the service delay experienced by the user, as $d_{i,j,t} = d'_{i,t} + d''_{i,j,t}$.

5.3.2 Workload Modeling

Workload characteristics along with other issues such as performance constraints determine the range of active servers needed in a cloud. Workload is usually characterized through statistical parameters of servers' traffic, those being average request arrival rate, peak arrival rate, service time and mean requests sizes. Since, the problem formulation accounts for migration overhead, we model workload at user level instead of request level. Note that migration overhead is due to moving state data associated with online users from source server to the destination server, thusly can be estimated from the number of online users whose application server is migrated. There are many research that verify the strong dependence between workload intensity and the number of online users [44]. Therefore, if one of those parameters is known, the other one can be estimated.

Let N_t be the set of the average numbers of online users in the areas for an epoch t, where $N_t = \{n_{1,t} \ldots n_{j,t} \ldots n_{|A|,t}\}$. The set N_t varies over time because, first, different applications have different local peak times during a day; second, the traffic peaks across the areas differ due to the time zone differences. We assume the population and distribution of the users and the electricity price to vary over time and space. However, we assume that these values remain constant within each epoch.

5.3.3 Energy Costs

We assume that the energy cost of a data center hosting a Web application is a function of the power model of its servers, its cooling energy and the electricity price.

We model the power consumption of a server at data center i at epoch t as: $p_{i,t} = u_{i,t} p_i^{util} + p_i^{idle}$, where p_i^{idle} is the per-server average idle power consumption for that data center, p_t^{util} is the additional power consumption of a server at full utilization with respect to idle, and $u_{i,t}$ is the utilization of the server at epoch t.

The utilization of a server depends on its workload and its physical characteristics. The workload of a server is a function of its online users. Therefore, we assume the following linear model for utilization of a server: $u_{i,t} = c_i n_{i,t}$, where c_i is the average utilization that one online user imposes on the server, and $n_{i,t} = \sum_{j=1}^{|A|} x_{i,j,t} n_{j,t}$ is the total number of users that are assigned to the data center i at epoch t. This model is frequently used in existing literature and experimental results show its sufficiency [1, 10].

Usually, many servers are allocated to the application. Assume n_i^{th} to be the total number of users that a single server in data center i can afford, i.e., $c_i n_i^{th} = u_i^{th}$, then the data center's cumulative idle power consumption equals to $y_{i,t} p_i^{idle}$, where $y_{i,t}$ is the number of active servers, calculated as $\lceil n_{i,t} / n_i^{th} \rceil$.

A data center's total power equals to the sum of computing and non-computing equipment power consumption (e.g., cooling power), and can be estimated as the product of its PUE and computing power. There are many data center metrics that evaluate its overall energy efficiency with respect to its computing energy efficiency. We choose PUE since it captures the data center energy inefficiency of the non-computing equipment with respect to computing energy in a linear way.

$$p_{i,t}^{total} = \left(\sum_{j=1}^{|A|} x_{i,j,t} n_{j,t} c_i p_{i,t}^{util} + y_{i,t} p_i^{idle} \right) PUE_i. \tag{5.1}$$

5.3.3.1 Total Energy Cost Considering Renewable Energy

To capture the effect of integrating renewable energy, we model energy buffering. According to this model flow of the renewable energy is smoothened using battery storage. Further, to maximize the renewable energy utilization over the energy drawn from the grid, the cost of renewables is set to *zero*. Let $r_{i,t}^{total}$ be the available renewable power at data center i which can be drawn at a time, $B_{i,t}$ denote the available renewable power at the battery, and B^{size} be the battery size to store and smooth renewables, the renewable harvesting at a time, $r_{i,t}$, always satisfies the following:

$$B_{i,t} - p_{i,t}^{AC} \tau + p_{i,t}^{totall} \tau + r_{i,t} \tau = B_{i,t+1}, B_{i,t} \leq B^{size}, \text{ and } r_{i,t} \leq r_{i,t}^{total}, \tag{5.2}$$

where, $p_{i,t}^{AC}$ denotes the total power draw from AC. If the total renewable power is always large enough, i.e., $r_{i,t} + B_{i,t} > p_{i,t}^{total}$, $\forall t$, data centers are called to be *sustainable*. While this is true if one can do perfect smoothing of the renewables using large-size storage, it is not always practically possible. This is due to, currently renewable power draw is a very small fraction of a data center's total power draw, there is energy leakage associated with storage, and there is size constraint for batteries because of battery costs and space limits of data centers. Considering zero cost for renewables, the total energy cost of a data center in an epoch can be calculated by multiplying the total energy draw from AC into the electricity price. We denote the electricity price by $e_{i,t}$, thus the total energy cost of an application hosted in data center i during epoch t equals to:

$$\text{cost}_{i,t}^{\text{energy}} = (p_{i,t}^{AC})\tau e_{i,t}. \tag{5.3}$$

5.3.4 Migration Cost

Dynamic workload distribution for stateful applications may require live migration (i.e., online users' state information should migrate from the source to the destination data center). Migration imposes a cost in terms of increase in network bandwidth consumption, and delaying the service for the affected online users. Therefore, we consider a uniform, per-user migration cost β, assuming equal-sized state information for all users. The calculation of the migration cost is based on the number of online users who have been migrated, as follows. Equation 5.3 suggests that if a front-end assignment to a data center between two intervals changes, then migration is performed. Therefore, we can calculate the number of migrated users for each data center and front-end by calculating the difference in the number of assigned users between two consecutive epochs. However, we choose not to directly take the difference between the previous epoch's $(t-1)$ assignment and the next epoch's (t) assignment, i.e., $n_{j,t}x_{i,j,t} - n_{j,t-1}x_{i,j,t-1}$, because we have to account for the users that are signing out in epoch $t-1$ (and therefore their connections are not migrated) and the users that are signing in, in epoch t (and therefore their connections did not exist at migration time). Let si denote the average fraction $(0 \leq si \leq 1)$ of new users out of the total users at each area over epochs, and so denote the average fraction $(0 \leq so \leq 1)$ of users at each area who sign out during each epoch, then the migration cost for a data center i at time t can be formulated as

$$\text{cost}_{i,t}^{\text{migration}} = \beta \sum_{j=1}^{|A|} \left((1 - si)n_{j,t}x_{i,j,t} - (1 - so)n_{j,t-1}x_{i,j,t-1} \right)^{+}. \tag{5.4}$$

Each of the si and so parameters can be estimated from the other based on preservation of flow, expressed by this relation: $n_{j,t}(1 - si) = n_{j,t-1}(1 - so)$ (i.e. the users that did not sign out in epoch $t-1$ should be equal to the online users that did not just sign in, in the epoch t). The decision to migrate workload is justified by the expectation that it can complete its execution by posing lower energy cost on another data center. The migration depends on two parameters: (i) the longevity of user connection; naturally, it is rarely beneficial to migrate a short running job as the benefit does not outweigh the migration costs; and (ii) the migration cost; if the migration cost is much higher than the difference between

Minimize

$$\text{Cost} = \text{Cost}^{\text{energy}} + \text{Cost}^{\text{migration}}$$

$$= \Sigma_{t=1}^{T} \left(p_{i,t}^{AC} \tau e_{i,t} + \beta \Sigma_{i=1}^{|S|} \Sigma_{j=1}^{|A|} \left((1-si)n_{j,t}x_{i,j,t} - (1-so)n_{j,t-1}x_{i,j,t-1} \right)^{+} \right), \quad (5.5)$$

subject to

$$(\text{Power constraint}) \; \forall i,t : p_{i,t}^{total} = (\sum_{j=1}^{|A|} x_{i,j,t}n_{j,t}c_i p_{i,t}^{util} + y_{i,t}p_i^{idle})PUE_i, \quad (5.6)$$

$$(\text{Buffering constraint}) \; \forall i,t : B_{i,t} + p_{i,t}^{AC} - p_{i,t}^{total} + r_{i,t} = B_{i,t+1}, \text{ and, } B_{i,t} \le B^{size}, \quad (5.7)$$

$$(\text{Service constraint}) \; \forall i,j,t : \; 0 \le x_{i,j,t} \le 1 \text{ and, } \sum_{i=1}^{|S|} x_{i,j,t} = 1, \quad (5.8)$$

$$(\text{Idle power constraint}) \; \forall i,j,t : \; y_{i,t} \in \mathbb{N}_0 \text{ and, } y_{i,t} \ge \frac{\sum_{j=1}^{|A|} x_{i,j,t}n_{j,t}}{n_i^{th}}, \quad (5.9)$$

$$(\text{Capacity constraint}) \; \forall i,t : \; 0 \le y_{i,t} \le s_{i,t}, \quad (5.10)$$

$$(\text{Performance constraint}) \; \forall i,j,t : \; d_{i,j,t} = d_i^{\text{ref}} + d_{i,j,t}'' \text{ and, } (d^{\text{ref}} - d_{i,j,t})x_{i,j,t} \ge 0. \quad (5.11)$$

Fig. 5.3 Mixed Integer Programming (MIP) formulation of GLP problem for interactive applications

energy cost efficiency of two data centers for processing an online user workload, the migration never happens. If the migration cost is much lower than the difference between energy-cost efficiency of two data centers, it always happens.

5.3.5 A Problem Formulation of GLP

The problem can be summarized as follows:

GLP problem: Given an application with a specific delay requirement d^{ref}, a cloud S_t in which the application can be hosted in a dynamic way, a spatio-temporal variation of the electricity price, $e_{i,t}$, a spatio-temporal variation of the number of the online users N_t, time-varying and fluctuating nature of renewable energy sources at each data centers, $r_{i,t}^{total}$, and size-constrained batteries to buffer low-cost energy, find the hosting for each epoch t that minimizes the sum of energy and migration cost, Eqs. 5.3 and 5.4.

All aforementioned costs are assumed to be monetary. We can model the application hosting problem as an optimization problem where the objective is minimizing the total cost as shown in Fig. 5.3.
Cost minimization is subject to the following constraints:

- *Power constraint* (Eq. 5.6), which states the total power required for computing and cooling at each data center.

- *Buffering constraint* (Eq. 5.7) which asserts the sufficiency of the power draw from battery and AC for the total required power, as well as the battery power level at a time.
- *Service constraint* (Eq. 5.8), which asserts that all users of every area should be assigned to a data center, and that there are no double assignments in either direction.
- *Idle power constraint* (Eq. 5.9), which ensures that the idle power consumption of all active servers is accounted.
- *Capacity constraint* (Eq. 5.10), which states that the number of assigned active servers to the application in a data center should not exceed the available servers (denoted by $s_{i,t}$) in that data center.
- *Performance constraint* (Eq. 5.11), which states that the traffic of end users should be split among data centers whose network and service delay is less than the users' delay requirement.

A solution to this problem would specify, at each epoch, how many servers in each data center should be assigned to the application (i.e., $y_{i,t}$), what portion of each area's traffic should be assigned to which data center (i.e., $x_{i,j,t}$), and how much is average power draw from AC, renewable and battery. Observe that some of the variables are reals (i.e., $x_{i,j,t}$, $p_{i,t}^{AC}$, and $B_{i,t}$) and some are integers (i.e., $y_{i,t}$). Therefore, due to linearity of all equations (both the objective function and the constraints), the problem is a Mixed Integer Programming, (MIP). MIP is a well-known NP-hard problem.

5.4 GLP Technical Challenges

5.4.1 Algorithmic Issues

The complexity of GLP solution is twofold. First, if we assume all information about workload, electricity price, renewable power over time are accurately available for a long decision period (i.e., $t = 0 \ldots T$), still the optimal solution can not be found in a time-efficient way using existing techniques. This is due to the NP-hardness of the problem: special case of the problem with zero migration cost and zero sized battery, is NP-hard [3]. In fact, under these assumptions, GLP can be formulated as a Fixed Charge Min Cost Flow (FCMCF) which is also a NP-hard [26] problem.

Second, in practice, it is quite often impossible to accurately know all of the information about electricity price, workload, and renewables in advance. In other words, the optimal solution, if any, can only be found offline. Therefore, to dynamically decide on the workload distribution across data centers, we need an online algorithm that competently decides on hosting of applications and workload placement. To ensure the performance of online algorithm compared to offline algorithm a theoretical competitive ratio is required which is usually not easy to calculate.

5.4.2 Prediction Issues

In practice, workload, renewable energy, and electricity price should be predicted over a window of epoch intervals to feed GLP. It is shown that the efficiency of GLP online solution compared to the optimal offline solution depends on the prediction window length [31]. The longer the prediction window is, the higher the performance of GLP online solution becomes compared to the optimal offline solution. While some of information, i.e., workload, electricity price, solar energy and are shown to have nice cyclic behavior, and thusly predictable, the others (i.e., wind energy) do not exhibit cyclic behavior and are thusly hard to predict.

5.4.3 General Model of GLP

In addition to the aforementioned challenges that deal with solving the given GLP problem, the formulation itself is lacking in some practical aspects.

First, the formulation assumes a single application type. In practice, data centers host different applications. In this regard GLP consolidate different applications over the cloud to incur minimum energy cost. Applications can be assigned to VMs, which can be placed at the most cost efficient data center at the time. However, consolidation of different set of applications comes with interference amongst them. Recent works suggest that consolidation of applications in a single server increases the contention on the shared resources such as on-chip caches, buses, main memory, CPUs and network [35, 36]. This contention results in performance degradation of applications. The performance overhead due to contention depends on the workload type of applications. The contention can also cause energy consumption overhead due to increase in runtime. Modeling such an effect and incorporating into GLP is not easy, since the interference effect depends on the workload type of applications and the workload intensity which are not easy to quantify and model for the scale of data centers [35].

Second, energy management in data centers is a cyber-physical problem in the sense that consolidating the workload on fewer servers affects the thermal conditions of data centers (e.g., creating hot spots due to the high power consumption of the active servers). However, energy formulation of GLP is given as a cyber problem and ignores thermal awareness into its formulation. The reason is the high cost that thermal awareness imposes on the solution, its evaluation, and its implementation (e.g. data center thermal modeling [46], managing a complex nonlinear cooling energy [48]). However recent studies show that depending on the power proportionality of servers, and thermal conditions of data center room, non-thermal aware server consolidation not only cannot reduce the energy consumption, but also increases the total energy consumption compared to no consolidation [4, 13] depending on data center cooling and computing power efficiency. This means that the current GLP formulation may not perform optimal server consolidation.

Third, GLP, as given, deals with steady state of data centers' dynamics i.e., workload, renewable energy sources, and thermal conditions (if thermal awareness would be incorporated in the current formulation) over an epoch. However, all of these parameters exhibit temporal fluctuation. Ignoring such temporal aspects can potentially affect the energy savings projected by GLP, since the decision making is performed according to inaccurate information. Due to management overhead of GLP, it is infeasible to choose very short decision time interval. However, initial transient analysis of GLP can help to optimally choose GLP parameters such as epoch length and prediction window to minimize such problems.

Finally, as discussed in Sect. 5.2.3 the physical limitations of batteries should be incorporated in GLP modeling, as they are significant factors [15, 51].

5.5 Existing Solutions and Related Work

This section reviews the existing work related to GLP problem, and the way that previous research resolve some of the challenges.

5.5.1 Proof of Concept: Trace Based Simulation Using Realistic Data

Qureshi et al. did the very first work in the area of workload management across data centers, to prove the concept and show effective parameters in the cost efficiency of the problem [40, 41]. The authors use heuristics to quantify the potential economic gain of considering electricity price in the location of computation. Through simulation using historical electricity prices, for twenty nine locations in the US, and network traffic data collected on Akamai CDN, they report that judicious location of computation load may save millions of dollars on the total operation cost of data centers. They also show that the magnitude of cost savings depends on how power-proportional the servers are and whether there is a constraint on the network bandwidth. They find that the cost saving is the highest when servers are ideally power-proportional and when the available network bandwidth is unconstrained. The results in this work reveals the potential gain of workload management across data centers for large providers, and some of the essential system requirements in practice, e.g., network bandwidth constraint.

5.5.2 Workload and Server Management for Stateless Applications ($\beta = 0$)

Considerable amount of research has been recently performed to find efficient GLP solutions for stateless workloads [2, 3, 31, 33, 42, 43, 52].

Le et al. developed a workload scheduling scheme across data centers for stateless applications where the problem is modeled as nonlinear optimization and it is solved using a meta-heuristic, i.e., *Simulated Annealing* [29]. The problem accounts for workload and server management across data centers, where the non-power-proportionality assumption of servers contributes in the nonlinearity of the problem. Their simulation results showed that by leveraging the electricity price, significant cost can be saved when servers are ideally power-proportional, and the cost saving decreases when servers have greater-than-zero idle power.

Rao et al. considered the load distribution of stateless applications across data centers with the objective of minimizing current energy cost subject to delay constraints [43]. The energy cost considered accounted for the average energy cost of active servers (i.e., active servers are assumed to operate at an average utilization and frequency). The authors used linear programming techniques and min-cost flow model to find an approximate solution. Abbasi et al. additionally enhance the power consumption model of active servers to be dependent on their current utilization and provide theoretical and numerical analysis of the approximation compared to the optimal solution [3]. The authors showed that the workload and server management for stateless applications can be modeled as fixed-charge min-cost flow problem, and find an N-approximation algorithm to solve it, where N is the number of class of servers across data centers. Rao et al. extended their scheme above (i.e., [43]) by developing a joint optimization of server management (i.e. resizing the active server set) and power management (i.e., CPU dynamic voltage and frequency scaling) across data centers using *General Benders Decomposition* [42]. In this model, DVFS technique is applied on active servers to reduce the processors' power consumption by scaling their frequency according to their offered workload.

Liu et al. tackled the management overhead of GLP by developing two distributed algorithms for achieving optimal geographical load balancing [33]. The authors develop a convex cost model which accounts for per active server energy cost, and delay cost. The delay cost is incurred due to overloading servers' or network propagation delay. They design decentralized algorithms which allow each data center and front-ends to optimize based on partial information. The authors provide theories to guarantee the convergence of algorithms solution to the optimal solution. The result in this work is very important, since the proposed algorithms simplify the implementation of GLP without degrading its performance.

Xu and Liu have developed GLP when jobs are a mix of delay-sensitive applications and delay-tolerant jobs, such as background/maintenance jobs [52]. The proposed solutions use delay-tolerant jobs to fill the extra capacity of data centers, give a higher priority to delay-sensitive jobs, and achieve good delay performance. Two algorithms are designed. (1) A stochastic subgradient-based algorithm, which solves a convex optimization problem for capacity allocation and load shifting in each slot. The solution is shown to converge to optimal cost. (2) A queue-based which also solves a convex optimization problem for capacity allocation and load shifting in each slot. The authors show this algorithm achieves optimal trade-off between queuing delay and cost.

5.5.3 Workload Management for Stateful Applications ($\beta \neq 0$)

Workload management for stateful applications does not receive much attention in the literature.

Buchbinder et al. developed a scheme for online job migration across data centers to reduce the electricity bill [8]. They assume data-intensive jobs for which migration causes overhead in terms of network delay and power. The authors make some simplification assumptions to provide theoretical results. They assume the migration cost is constant over all jobs, and more importantly they assume the total load over time is constant. To optimize the electricity cost, the authors develop a cost model consisting of server' electricity cost and job migration. Since migration cost depends on every two consecutive job assignment, the optimal solution can only be found offline. Authors design an online algorithm and prove that it has a competitive bound of $\log(n)$ compared to the offline optimal solution, where n is the total number of servers across the cloud. However, due to the complexity of the algorithm, an easy-to-implement heuristic online algorithm is proposed which is evaluated through simulation using real electricity pricing and job workload data. The assumptions to derive the analytical bound was more suited to batch jobs.

Abbasi et al., modeled the GLP for Web based stateful applications jobs. The migration cost is modeled as a function of the number of migrated user connections (see Sect. 5.3.4). Through the simulation study, the authors argue that when migration cost is comparable to the *reference energy-cost benefit of a migration*, i.e., the average energy-cost difference between data centers, migration does not drop the cost efficiency of GLP significantly.

5.5.4 Renewable Energy Utilization Within and Across Data Centers

Some related work propose green scheduling algorithms to maximally utilize renewable [5, 32, 34, 45, 53]. The idea is to adjust the power consumption to the available green power supply using power management techniques, e.g., server power state transitions and workload shifting. Liu et al., perform a numerical study to evaluate the utilization of renewable energy with size constrained batteries across data centers. Their simulation study highlights the efficiency of "follow the renewable" strategy [32]. Further, they show that small-size batteries can be beneficiary to efficiently utilizing renewables. The authors further perform a theoretical and empirical study in a HP data center to jointly optimize renewable and cooling energy within a data center [34]. Zhang et al., propose GreenWare for Internet data centers to maximize the percentage of renewable energy used to power a network of distributed data centers,subject to the desired cost budget [53]. Finally, [45] focuses on workload distribution based on renewable availability and energy cost across a set of distributed data centers.

5.5.5 Energy Buffering Management

Recently, energy buffering to exploit low-cost electricity has drawn attention [15, 47]. Data centers are conventionally equipped with UPSes for emergency case. Govindan et al. and Urgaonkar et al. propose to partially utilize UPSes for energy cost management [15, 47]. The idea is to store energy in UPS batteries during "valleys" periods of lower demand, which can be drained during "peaks" periods of higher demand. Urgaonkar et al. develop an on-line control algorithm using Lyponov to exploit UPS devices across data centers [47], and Govindan et al. perform a comprehensive study on the feasibility of utilizing UPS to store low-cost energy, the constraints (e.g., charging discharging periods depending on life-cycle of batteries) and a Markovian based solution to [15].

Wang et al., investigate how data centers can leverage the existing huge set of heterogeneous Energy Storage Devices (ESDs) [51]. The authors argue that continuing technology advances provides a plethora of competitive ESD options which offer different trade-offs between lifetime, energy efficiency, and cost. Further, ESD devices can be placed in different levels of data centers power hierarchy (i.e., data center, rack, and server levels). The authors developed useful cost models to study the cost-benefit of various ESDs. They also presents a theoretical framework to quantify the cost-benefit trade-offs of various ESD options as a function of workload properties.

5.5.6 Online Algorithms Versus Offline Algorithms to Manage Energy Buffering and Server Switching GLP

There have been few works which try to design online algorithms with guaranteed competitive bound compared to the optimal offline algorithms. Lin et al., propose online algorithms for GLP when there is a cost associated to switching servers to on [31]. Since server switching cost depends on every two consecutive active server set, the GLP solutions over time are dependent. Therefore, the optimal solution can only be found offline. The authors evaluate a commonly used algorithm for GLP which suggest to perform workload management for the current time by optimizing cost over a time window where load is predicted. They show that such an algorithm performs well compared to the offline optimal provided that servers are homogeneous. Particularly, the competitive bound decreases with increasing prediction window length. They propose a new online algorithm for the case of heterogeneous servers with guaranteed competitive bound of $(1 + O(\frac{1}{w}))$, where w denotes the prediction window length.

Another online algorithm for GLP is proposed in [47]. Urganokar et al., consider the problem of using UPS batteries to reduce the time average electric utility bill in a data center. The problem accounts for the battery energy management to shift the peak demand away from high tariff period. Using the technique of Lyapunov

Table 5.2 A taxonomy of GLP existing solutions

Applications' types	Management	Articles
Stateless interactive jobs	Workload and server management	[2, 3, 29, 31, 33, 40–43, 52]
	Workload, server, and energy buffering	[32, 34, 47]
	Energy buffering	[15, 51]
	Renewable harvesting	[5, 32, 34, 45, 53]
Stateful interactive jobs	Workload and server management	[2, 3]
Stateful batch jobs	Workload and server management	[8]
Interactive and batch jobs	Workload and server management	[52]

optimization, the authors develop an online control algorithm that can optimally exploit batteries to minimize the average cost. Interestingly, this algorithm does not need any knowledge of the statistics of the workload or electricity cost processes. The authors also show that the deviation of the algorithm from the optimal solution reduces as the storage capacity is increased.

5.5.7 Summary

The above work highlights that GLP problem for stateless applications can be approximately modeled by a linear or a convex cost function, and that the approximation yields negligible ratio/effect in its solution. GLP discretizes two way connection between proxies and data centers, and yield optimal solution. This has much simpler implementation with less network and management overhead compared to a centralized solution.

Existing work lacks the accurate migration modeling of jobs as well as efficient online algorithms which account for migration cost. The bandwidth costs associated with moving the applications might be a significant factor and might vary over cloud. Further the migration might affect the application's delay, and thusly violate the performance requirement of delay-sensitive applications. Quantifying such parameters to accomplish joint optimization of power cost and migration cost is a challenge. Empirical studies can help to resolve such challenges (Table 5.2).

Despite progress on developing efficient GLP algorithms for stateless applications, many aspect of the problem has not been addressed sufficiently.

Specially, the performance of GLP depends on the predictability of information such as workload, and renewable energy sources, however there is no work which suggest what prediction technique and configuration (i.e. prediction time interval and window length) incur the lowest cost.

Data centers host different applications ranging from batch jobs [35] to web services (e.g., e-commerce). The workload types of such applications, e.g., CPU-intensive workloads (e.g., to analyze or organize data) and memory-intensive workloads (e.g., data retrieval tasks), and their performance requirements (e.g., delay-sensitive and delay -tolerant applications) are different. To achieve sustainability, GLP should account for a coordinated management of set of heterogeneous applications, as well as batteries and workload assignment to adapt the total computing over all data centers in the cloud to the total low-cost energy supply.

Moreover, all of the above work only account for a cyber-model of power consumption, ignore temporal dynamics of data center workload, batteries and thermal condition, and only use a trace-based evaluation.

5.6 Evaluating the Efficiency of GLP for Developing Sustainable Data Centers

We perform a trace-based simulation to evaluate the efficiency of GLP using realistic renewable energy profiles and workload traces.

5.6.1 Simulation Setup

We simulate a cloud consisting of three data centers. Their workload, electricity and physical characteristics, such as server power profiles, battery are set according to realistic data. To this end, we assume data centers are located at the following three locations: Atlanta, GA; Houston, TX; and Mountain View, CA, namely DC1, DC2 and DC3, respectively. These locations correspond to the location of three major Google data centers. We used the historical electricity prices for the above locations [43] (see Fig. 5.1a). Note that, in reality, each data center provider may have different electricity price contracts, i.e., lower electricity price than households. The electricity price of Fig. 5.1a is used as an *example* to show the cost saving benefit of GLP by leveraging electricity cost. We also assume a battery for each data centers where its size vary between 0 and 6,400 MJ (this size is equal to the average energy consumption of data centers in 24 h).

5.6.1.1 Data Center Types

Three homogeneous (identical) data centers are considered for the simulation with contemporary servers (e.g., IBM Systems x3650 M2: idle power 100 and peak power 320 W). The maximum number of servers for each data center is set to 400 which matches the workload intensity range used in the simulations.

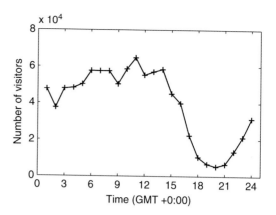

Fig. 5.4 Hourly number of U.S. online users for an entertainment Web site hosted at GoDaddy.com on 17th March, 2011 [3]

To model the utilization of servers, we assume that each online user imposes 0.0005 utilization to each server (i.e., $c = 0.0005$) in Eq. 5.9 from Fig. 5.3 and for the sake of simplicity we assume servers can be utilized up to 100% and there is no network delay.

5.6.1.2 Workload Distribution

We used one day (March 17, 2011) of workload trace of an entertainment Web site hosted at *GoDaddy.com*. Using *Google Analytics*, we collected the hourly total number of visitors to the Web site from different USA states (see Fig. 5.4). The workload is scaled up to the data centers' capacity.

5.6.1.3 Renewable Energy Profile

We use realistic traces of wind and solar energy from [22, 23] that have measurements every 10 min for a year across different location in USA. We choose two centers in CA and TX to capture the availability of solar and wind energy in DC1 and DC2. Since we did not find any center at GA, we choose a renewable profile of a center in IL instead to use at DC3 (see Fig. 5.1c, d).

5.6.1.4 Experiments Performed

We performed different experiments to show how GLP along with energy buffering management can push energy/cost sustainability through increasing the utilization of renewable and decreasing electricity cost. We evaluate the cost efficiency of GLP with respect to battery sizes of data centers, and energy renewable prediction window length. We run two workload and server management algorithms, (i) GLP

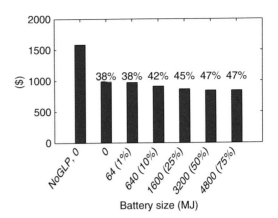

Fig. 5.5 Electricity cost of GLP versus NoGLP and battery sizes

that is the solution of the problem shown in Fig. 5.3 (with a real value assumption for the variable y), and (ii) NoGLP where workload is statically balanced among data centers and no dynamic workload shifting across data centers is allowed; to solve NoGLP, the problem in Fig. 5.3 is independently solved for each data center (with a real value assumption for the variable y)).

We used *GNU Linear Programming Kit (GLPK)* solver under MATLAB 2009, to solve GLP and noGLP algorithms.

5.6.2 GLP Electricity Cost Saving

In this experiment, we compare the GLP electricity cost to NoGLP cost with respect to battery size. We vary the battery size form 0 to 6,400 MJ, where 6,400 MJ is equal to the average energy consumption of data centers in 24 h. We scale renewable traces such that when GLP runs with zero battery size the average renewable percentage equals to 5%. Results in Fig. 5.5 show that GLP saves 38% electricity cost compared to NoGLP when there is no battery. Saving increases up to 47% with increasing battery. The reason is that GLP utilizes batteries to store low cost energy during low-electricity-rate periods.

5.6.3 Sustainability Versus GLP and Battery Size

In the following experiment, GLP along with energy buffering helps to increase the sustainability of data centers through increasing renewable energy utilization. We run the experiment for several values of battery sizes ranging from 0 upto 6,400 MJ. Finally, we compare the optimal offline solution of both GLP and NoGLP for a time horizon of 1 week ($T = 7 \times 24$), where all of the information of renewable and workload for the entire time horizon is available.

Fig. 5.6 Renewable energy utilization versus battery size

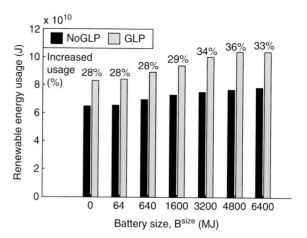

Fig. 5.7 Energy consumption (type/magnitude) of GLP and NoGLP versus different battery sizes (bars are sorted according to battery sizes of Fig. 5.6)

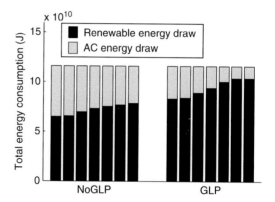

We scaled the renewable linearly for all data centers, and run both GLP and NoGLP. Results shown in Fig. 5.6 indicates that GLP utilizes renewable 28% more than NoGLP when there is no battery. Note that renewable energy profile for each data centers in both GLP and NoGLP are the same, however GLP can utilize it more than NoGLP under the same battery size.

With increasing battery size, both GLP and NoGLP increase the utilization of renewable. However, to achieve the same renewable energy usage, NoGLP needs much larger battery size compared to GLP.

Figure 5.7 showing total energy usage (type and magnitude), indicates how GLP can push sustainability in data centers using both workload placement across data centers and energy buffering management.

Figures 5.8 and 5.9 show the energy consumption and energy type of all three data centers using GLP and NoGLP over time. The figures shows that NoGLP buffers renewable energy (or low-cost electricity price) during low workload time

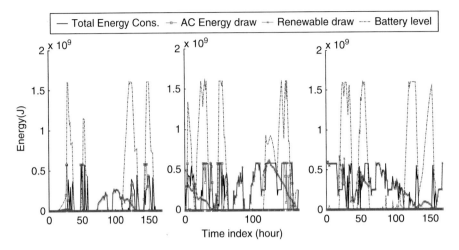

Fig. 5.8 Energy consumption across all data centers using GLP and battery size of 1,600 MJ

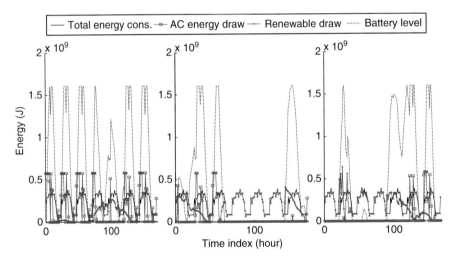

Fig. 5.9 Energy consumption across all data centers using NoGLP and battery size of 1,600 MJ

(low utility rate) and utilizes it during peak workload time (high utility rate).
Along with this, GLP also manages the workload share of each data centers to
minimize cost.

5.6.4 Workload and Renewable Energy Prediction

In this section, we use different time series prediction techniques to investigate the
predictability of renewable energy sources and workload. We use solar and wind

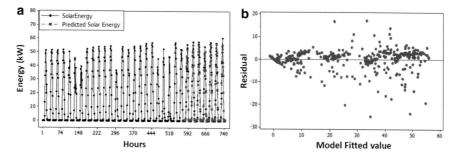

Fig. 5.10 (a) The solar energy trace indicates daily seasonal behavior which is captured with the with SARIMA(002,111). (b) The model fitted indicates that the residual values are close to 0 for most of the fitted values

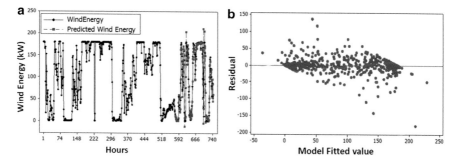

Fig. 5.11 (a) The wind energy trace indicates no seasonal behavior hence simple moving average model can be used. (b) The model fitted indicates that the residual values are close to 0 for most of the fitted values

energy traces obtained from the sites in California, Texas and Illinois [22, 23] (see Fig. 5.1c, d).

We used different techniques such as the moving average based technique of Holt Winters, auto regression techniques (i.e., linear, linear with seasonal, linear with harmonics), and SARIMA (Seasonal Auto Regressive Integrated Moving Average).

We found that to model the cyclic behavior in the solar energy trace, the SARIMA model performs better than the other models (see Fig. 5.10a) where as a simple moving average model is better for wind traces (see Fig. 5.11a).

The accuracy of the models is decided based on several factors such as the residual values obtained at different lags, the fitting of the residual value as in Figs. 5.10b and 5.11b. We also use the Auto Correlation Function and Partial Auto Correlation plots of the residuals to make sure that residuals do not exhibit any correlation.

To evaluate the predictability of solar and wind energy traces, we calculate the absolute prediction error and normalize it to compare the predictability of the two traces.

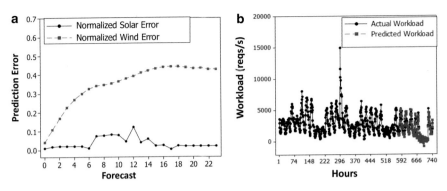

Fig. 5.12 (**a**) Error for different forecast window for wind and solar, (**b**) prediction of NASA workload trace using SARIMA

As shown in Fig. 5.12a, the prediction error of the wind trace increases rapidly with the size of the forecast window. The same figure shows that the prediction error of solar energy remains almost constant with the size of the forecast window.

We also investigate the predictability of the Internet workload using NASA traces. We develop an ARIMA based prediction model and observe the absolute prediction error to be 5.8% over a forecasting window length of 24.

5.6.5   Sustainability Versus Prediction Window

The GLP evaluation results that is presented so far were based on an GLP offline solution which is aware of the information of the entire time horizon in advance. The results in the previous section highlights that the information such as wind traces can not be accurately predicted.

In this experiment we solve GLP for a prediction window of size 24, and compare the results with the offline solution results where GLP is solved for the entire time horizon of 168 (a week). Results as shown in Fig. 5.13 indicates of up to 8% reduction in renewable energy usage of GLP compared to the offline solution.

5.6.6   Discussion on the Results

The simulation study highlights the usefulness of GLP in reducing electricity cost, and its potential aid in reducing energy infrastructure cost to achieve sustainability (e.g., renewable energy generation and energy storage). The results show that while GLP achieves 75% energy sustainability, i.e. the contribution of renewable energy in the total energy consumption, with average renewable generation rate of 54 MWh and battery size of 640 MJ, No GLP achieves 60% sustainability under

Fig. 5.13 Renewable energy versus prediction window length

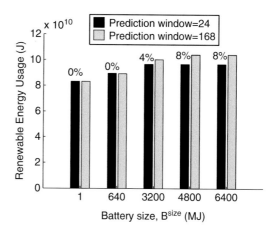

the same condition of renewable energy generation and battery size (refer Fig. 5.6). The simulation however, is performed under ideal assumptions of zero bandwidth cost, zero migration cost, and ideal batteries (e.g., 100% energy efficiency). In practice the benefit of GLP may come at the cost of using high cost bandwidth, and consuming higher bandwidth than NoGLP (the extra bandwidth usage is due to migration of users'/applications' state information). Further, the physical limitation of batteries prevent GLP to charge/discharge battery at any frequency, time and rate. Future work should address the cost-benefit of GLP in developing sustainable data centers under physical characteristics of batteries and bandwidth costs.

The performance of GLP is also affected by predictability of information including workloads and renewable. The longer prediction window length increases the cost efficiency of GLP. However, our results show that the prediction error of wind rapidly increases with increasing prediction window. This necessitates to design efficient online algorithms with small competitive ratio with respect to the offline optimal solutions than runs on a long time horizon.

Finally, the practical issues of GLP should be tested and examined using data center research infrastructure such as the BlueTool [17, 19], which offers a small data center for experimentation with innovative management schemes.

5.7 Conclusions

This study gave an overview on Geographical appLication/WorkLoad Placement (GLP) which has recently drawn attention in research and industry. Existing research highlights its usefulness in increasing the use of renewable energy by managing the costs of energy buffering and bandwidth costs. Despite all this, a lot still needs to be done. Particularly, there is no empirical study to expose the practical challenges of GLP. Further, the practical issues to migrate stateful applications' workload, heterogeneity of applications, joint optimization of computing and cooling energy across data centers has not been sufficiently addressed.

This chapter also presented a general cost model of GLP for interactive stateful applications that accounts for minimizing electricity cost price and maximizing renewable energy usage across data centers. The model along with realistic workload and renewable traces used in a simulation study to evaluate the efficiency of GLP in achieving sustainable data centers. Results show that, GLP, can indeed decreases the need for large-scale batteries to move toward sustainable data centers. This can be achieved if predicted workload and renewable energy information is provided with reasonably accuracy.

References

1. Abbasi Z, Varsamopoulos G, Gupta SKS (2010) Thermal aware server provisioning and workload distribution for Internet data centers. In: ACM international symposium on high performance distributed computing (HPDC10), Chicago, pp 130–141
2. Abbasi Z, Mukherjee T, Varsamopoulos G, Gupta SKS (2011) Dynamic hosting management of web based applications over clouds. In: International conference on high performance computing conference (HiPC2011), Bengaluru
3. Abbasi Z, Mukherjee T, Varsamopoulos G, Gupta SKS (2012) DAHM: a green and dynamic web application hosting manager across geographically distributed data centers. J Emerg Technol Comput Syst 8(4):34:1–34:22
4. Abbasi Z, Varsamopoulos G, Gupta S (2012) Tradeoff and energy awareness in IDC management. ACM Trans Archit Code Optim 9(2):11
5. Akoush S, Sohan R, Rice A, Moore AW, Hopper A (2011) Free lunch: exploiting renewable energy for computing. In: Proceedings of HotOS, Napa
6. Barroso LA, Holzle Urs (2007) The case for energy-proportional computing. Computer 40:33–37
7. Beskow PB, Vik K-H, Halvorsen P, Griwodz C (2009) The partial migration of game state and dynamic server selection to reduce latency. Multimed Tools Appl 45(1):83–107
8. Buchbinder N, Jain N, Menache I (2011) Online job-migration for reducing the electricity bill in the cloud. Networking 6640:172–185
9. Barbir A., Cain B, Nair R, Spatscheck O (2003) Known content network (CN) request-routing mechanisms. Internet engineering task force RFC 3568
10. Chase J, Anderson D, Thakar P, Vahdat A, Doyle R (2001) Managing energy and server resources in hosting centers. In: Proceedings of the eighteenth ACM symposium on operating systems principles (SOSP 2001), Lake Louise, pp 103–116
11. Chen Y, Das A, Qin W, Sivasubramaniam A, Wang Q, Gautam N (2005) Managing server energy and operational costs in hosting centers. SIGMETRICS Perform Eval Rev 33(1): 303–314
12. Conti M, Gregori E, Panzieri F (2000) Load distribution among replicated web servers: a qos-based approach. ACM SIGMETRICS Perform Eval Rev 27(4):12–19
13. Faraz A, Vijaykumar TN (2010) Joint optimization of idle and cooling power in data centers while maintaining response time. ACM SIGARCH Comput Archit News 38(1):243–256
14. GSA FEMP (2010) Quick start guide to increase data center energy efficiency. Technical report, General Services Administration (GSA) and the Federal Energy Management Program (FEMP)
15. Govindan S, Sivasubramaniam A, Urgaonkar B (2011) Benefits and limitations of tapping into stored energy for datacenters. In: Proceedings of the 38th international symposium on computer architecture (ISCA), San Jose

16. Guoju Z, Xisheng T, Zhiping Q (2010) Research on battery supercapacitor hybrid storage and its application in microgrid. In: Power and energy engineering conference (APPEEC), 2010 Asia-Pacific, Chengdu, pp 1–4
17. Gupta SKS, Gilbert RR, Banerjee A, Abbasi Z, Mukherjee T, Varsamopoulos G (2011) GDCSim: a tool for analyzing green data center design and resource management techniques. In: Proceedings of international green computing conference (IGCC11), Orlando. IEEE
18. Gupta SKS, Mukherjee T, Varsamopoulos G, Banerjee A (2011) Research directions in energy-sustainable cyber-physical systems. Elsevier Sustain Comput (SUSCOM) 1(1):57–74. Invited paper
19. Gupta SKS, Varsamopoulos G, Haywood A, Phelan P, Mukherjee T (2012) BlueTool: using a computing systems research infrastructure tool to design and test green and sustainable data centers. In: Ahmad I, Ranka S (eds) Handbook of energy-aware and green computing, number 45. Chapman and Hall/CRC, Boca Raton
20. http://www.apple.com/environment/renewable-energy/
21. http://www.google.com/about/datacenters/renewable/index.html
22. http://rredc.nrel.gov/solar/new_data/confrrm/
23. http://wind.nrel.gov/Web_nrel/
24. Koomey JG, Belady C, Patterson M, Santos A, Lange K-D (2009) Assessing trends over time in performance, costs, and energy use for servers. Technical report, Microsoft Corporation and Intel Corporation
25. Krioukov A, Mohan P, Alspaugh S, Keys L, Culler D, Katz R (2010) NapSAC: design and implementation of a power-proportional web cluster. In: Proceedings of the first SIGCOMM workshop on green networking, New Delhi. ACM, pp 15–22
26. Krumke SO, Noltemeier H, Schwarz S, Wirth H-C, Ravi R (1999) Flow improvement and network flows with fixed costs. In: Operations research proceedings 1998, Zurich. Springer, Berlin/Heidelberg, pp 158–167
27. Kumar K, Lu Y-H (2010) Cloud computing for mobile users: can offloading computation save energy? Computer 99:51–56
28. Kusic D, Kephart JO, Hanson JE, Kandasamy N, Jiang G (2009) Power and performance management of virtualized computing environments via lookahead control. Clust Comput 12:1–15
29. Le K, Bilgir O, Bianchini R, Martonosi M, Nguyen TD (2010) Managing the cost, energy consumption, and carbon footprint of Internet services. SIGMETRICS Perform Eval Rev 38(1):357–358
30. Lin M, Wierman A, Andrew LLH, Thereska E (2011) Dynamic right-sizing for power-proportional data centers. In: Proceedings of the IEEE INFOCOM, Shanghai, pp 10–15
31. Lin M, Liu Z, Wierman A, Andrew LLH (2012) Online algorithms for geographical load balancing. In: Proceedings of international green computing conference (IGCC11), Orlando. IEEE
32. Liu Z, Lin M, Wierman A, Low SH, Andrew LLH (2011) Geographical load balancing with renewables. ACM SIGMETRICS Perform Eval Rev 39(3):62–66
33. Liu Z, Lin M, Wierman A, Low SH, Andrew LLH (2011) Greening geographical load balancing. In: Proceedings of the ACM SIGMETRICS, San Jose. ACM, pp 233–244
34. Liu Z, Chen Y, Bash C, Wierman A, Gmach D, Wang Z, Marwah M, Hyser C (2012) Renewable and cooling aware workload management for sustainable data centers. ACM SIGMETRICS Perform Eval Rev 40:175–186. ACM
35. Mars J, Tang L, Hundt R, Skadron K, Soffa ML (2011) Bubble-up: increasing utilization in modern warehouse scale computers via sensible co-locations. In: Proceedings of the 44th annual IEEE/ACM international symposium on microarchitecture, New York. Porto Alegre, Brazil ACM, pp 248–259
36. Merkel A, Stoess J, Bellosa F (2010) Resource-conscious scheduling for energy efficiency on multicore processors. In: Proceedings of the 5th European conference on computer systems, EuroSys '10. ACM, New York, pp 153–166
37. Miller R (2011) Facebook installs solar panels at new data center. White Paper

38. Mukherjee T, Banerjee A, Varsamopoulos G, Gupta SKS (2010) Model-driven coordinated management of data centers. Comput. Netw. 54(16):2869–2886
39. Pathan M, Vecchiola C, Buyya R (2008) Load and proximity aware request-redirection for dynamic load distribution in peering CDNs. In: On the move to meaningful internet systems: OTM 2008. Springer, Berlin/Heidelberg, pp. 62–81
40. Qureshi A (2010) Power-demand routing in massive geo-distributed systems. PhD thesis, Massachusetts Institute of Technology
41. Qureshi A, Weber R, Balakrishnan H, Guttag J, Maggs B (2009) Cutting the electric bill for Internet-scale systems. In: Proceedings ACM SIGCOMM, Barcelona, pp 123–134
42. Rao L, Liu X, Ilic M, Liu J (2010) MEC-IDC: joint load balancing and power control for distributed Internet data centers. In: Proceedings of the 1st ACM/IEEE international conference on cyber-physical systems, Stockholm, pp 188–197
43. Rao L, Liu X, Xie L, Liu W (2010) Minimizing electricity cost: optimization of distributed Internet data centers in a multi-electricity-market environment. In: Proceedings of the IEEE INFOCOM, San Diego, pp 1–9
44. Saroiu S, Gummadi KP, Dunn RJ, Gribble SD, Levy HM (2002) An analysis of internet content delivery systems. ACM SIGOPS Oper Syst Rev 36(si):315
45. Stewart C, Shen K (2009) Some joules are more precious than others: managing renewable energy in the datacenter. In: Workshop on power aware computing and systems, Big Sky
46. Tang Q, Gupta SKS, Varsamopoulos G (2008) Energy-efficient thermal-aware task scheduling for homogeneous high-performance computing data centers: a cyber-physical approach. IEEE Trans Parallel Distrib Syst 19(11):1458–1472
47. Urgaonkar R, Urgaonkar B, Neely MJ, Sivasubramanian A (2011) Optimal power cost management using stored energy in data centers. In: Proceedings of the ACM SIGMETRICS joint international conference on measurement and modeling of computer systems, San Jose, pp 221–232. ACM
48. Varsamopoulos G, Banerjee A, Gupta SKS (2009) Energy efficiency of thermal-aware job scheduling algorithms under various cooling models. In: International conference on contemporary computing IC^3, Noida, pp 568–580
49. Varsamopoulos G, Gupta SKS (2010) Energy proportionality and the future: metrics and directions. In: 39th international conference on parallel processing workshops (ICPPW), San Diego. IEEE, pp 461–467
50. Vokoun R (2012) Renewable energy in today's data center. White Paper
51. Wang D, Ren C, Sivasubramaniam A, Urgaonkar B, Fathy H (2012) Energy storage in datacenters: what, where, and how much? In: Proceedings of the 12th ACM SIGMETRICS/PERFORMANCE joint international conference on measurement and modeling of computer systems, London. ACM, pp 187–198
52. Xu D, Liu X (2012) Geographic trough filling for internet datacenters. In: INFOCOM, 2012 Proceedings IEEE, Orlando. IEEE, pp 2881–2885
53. Zhang Y, Wang Y, Wang X (2011) Greenware: greening cloud-scale data centers to maximize the use of renewable energy. In: Middleware 2011, Lisboa, Portugal, pp 143–164

Chapter 6
Barely Alive Servers: Greener Datacenters Through Memory-Accessible, Low-Power States

Vlasia Anagnostopoulou, Susmit Biswas, Heba Saadeldeen, Alan Savage, Ricardo Bianchini, Tao Yang, Diana Franklin, and Frederic T. Chong

6.1 Introduction

Energy represents a large fraction of the operational cost of Internet services. As a result, previous works have proposed approaches for conserving energy in these services, such as consolidating workloads into a subset of servers and turning others off [8–10, 31], and leveraging dynamic voltage and frequency scaling of the CPUs [9, 12, 13].

Consolidation is particularly attractive for two reasons. First, current resource provisioning schemes leave server utilizations under 50% almost all the time [13]. At these utilizations, server energy efficiency is very low [4]. Second, current servers consume a significant amount of energy even when they are completely idle [4]. Despite its benefits, services typically do not use this technique. A major reason is the fear of high response times during re-activation in handling traffic spikes. Another reason is that services often want the memory and/or storage of all servers to be readily available even during periods of light load. For example, interactive services try to maximize the amount of memory available for data caching across the cluster, thereby avoiding disk accesses or content re-generation.

In this paper, we propose an approach that does not completely shutdown idle servers, enables fast state transitions, and keeps in-memory application code/data untouched. Specifically, we propose to send servers to a new family of "barely-alive" power states, instead of turning them completely off after consolidation.

V. Anagnostopoulou (✉) • S. Biswas • H. Saadeldeen • A. Savage • T. Yang
D. Franklin • F.T. Chong
University of California Santa Barbara, Santa Barbara, CA 93106, USA
e-mail: vlasia@cs.ucsb.edu; susmit@cs.ucsb.edu; heba@cs.ucsb.edu; asavage@cs.ucsb.edu;
tyang@cs.ucsb.edu; franklin@cs.ucsb.edu; chong@cs.ucsb.edu

R. Bianchini
Rutgers University, Piscataway, NJ 08854, USA
e-mail: ricardob@cs.rutgers.edu

P.P. Pande et al. (eds.), *Design Technologies for Green and Sustainable Computing Systems*, 149
DOI 10.1007/978-1-4614-4975-1_6, © Springer Science+Business Media New York 2013

In a barely-alive state, a server's memory (and possibly its disks) can still be accessed, even if many of its other components are turned off. Keeping data active and accessible in barely-alive states enables software to implement cluster-wide (or "cooperative") main-memory caching, data replication and coherence, or even cluster-wide in-memory data structures, while conserving a significant amount of energy.

Our evaluation starts by comparing barely-alive states to conventional consolidation via complete server shutdown, as well as more recent proposals such as PowerNap and Somniloquy. In particular, we evaluate the effect of server restart latency on response time during typical load spikes. Spikes may occur due to a variety of reasons, including external events (e.g., Slashdot effect), the temporary unavailability of a mirror datacenter, operator mistakes, or software bugs. Under latency constraints, greater restart latency translates to a larger number of extra active servers provisioned to absorb the load. We evaluate the sensitivity of each energy conserving scheme to the duration and magnitude of load spikes, as well as to modifications to data while in energy-conserving server states.

We then present a study of a server cluster implementing a cooperative cache for the "snippet" generator of a Web search service. Many services today use cooperative caching middlewares (e.g., Memcached is used at Wikipedia, Twitter, and others [11]). Our cooperative caching implementation accommodates barely-alive servers and dynamically re-sizes the cache as a function of workload variations and desired performance. Any memory not used for caching can be used by other applications. For this study, we simulate systems based on an efficient barely-alive state, on-off, Somniloquy, and low-end servers. We also investigate the tradeoff between performance and energy savings under various system parameters.

Finally, we introduce two case studies using the barely-alive states. In the first study, we propose a "mixed" system that combines active, barely-alive and off states. We find that at each performance level, the mixed system achieves the highest energy savings. In the second study, we propose a system that combines active and barely-alive only, but hosts more than one service. Overall, barely-alive states can produce energy savings of up to 38%, compared to a baseline energy-oblivious system. Moreover, we find that barely-alive states can conserve significant energy across a large parameter space. When two services share the cluster, the barely-alive system can save up to 34% energy.

The remainder of the paper is organized as follows. Next, we discuss the background and right after the related work. In Sect. 6.4, we introduce the barely-alive family of power states. We qualitatively compare our family of states to previous schemes in Sect. 6.5. Section 6.6 presents our analysis of provisioning for load spikes. In this section, we also describe our simulation infrastructure and aggregate memory results. In Sect. 6.7 we introduce the mixed system case study and assess the energy savings of the mixed system at different performance levels as compared to other approaches. In Sect. 6.8 we introduce the memory sharing functionality of our caching middleware with the barely-alive system, and evaluate its potential for energy savings at the presence of two services. Finally, we draw our conclusions in Sect. 6.9.

6.2 Background

6.2.1 Consolidation and Low-Power Server States

The idea of dynamic workload consolidation consists of adjusting the number of active servers dynamically, based on the load offered to the service. During periods of less-than-peak load, the workload can be concentrated (either through state migration or request distribution) on a subset of the servers and others can be turned off. An alternative to turning servers off is to transition the rest of the servers into a low power state.

Two low-power states have been proposed recently. Somniloquy [2] augments the network interface to be able to turn most other components off during periods of idleness, while retaining network connectivity. In the low-power state, *main memory becomes inaccessible,* so accesses can only be performed to the small memory of the network interface. *No disk accesses can be effected.* Moreover, *updates to main memory can only be performed after activation,* thereby increasing delay. In contrast, our states allow read and write accesses to the entire main memory and disks. PowerNap rapidly [27] transitions servers between active and "nap" state, obviating the need for consolidation. *In nap state, a server is not operational.* PowerNap requires server software to avoid unwanted transitions to active state (e.g., due to clock interrupts). More challengingly, PowerNap requires the server to be *completely idle,* which is becoming harder as the number of cores per CPU increases (the idle times of all cores must perfectly overlap). Recently, Meisner and Wenisch addressed the latter problem by forcing idle times to overlap and adding a co-processor to each server [26].

6.2.2 Cooperative Caching

Cluster-wide cooperative main-memory caching (or simply cooperative caching) [6, 11, 30] improves the performance of Internet services, as it caches the most popular objects in the server memories, thereby avoiding disk accesses or content regeneration. The mapping of objects to server memories is known to the intra-cluster request distribution algorithm. However, existing cooperative caching layers differ in how they distribute the incoming client requests across the cluster. In the layer we study in this paper [6], when a client request arrives, the distribution algorithm directs it to one of the servers that caches the requested object (if one exists), as long as that does not excessively imbalance the load across servers. When it does, the caching layer creates an additional replica of the object to better spread the load.

An important characteristic of the caching layer is the object placement and replacement. A simple LRU cache has been found to yield good results [14] for this. At the same time, it is important to have a notion of the hit ratio of a cache hierarchy, given its capacity, the replacement policy and a sequence of memory accesses, in order to implement memory allocation and sharing policies for the applications. An interesting approach for LRU caches is the Stack algorithm [25], which can compute the hit ratio that would be achieved by all cache sizes using a single pass over the stream of memory accesses. The idea of the algorithm is to keep an "LRU stack" of memory block (e.g. main memory page) addresses sorted by recency of access; an access moves the corresponding block address to the top of the stack. In addition, the algorithm computes the "stack distance" between two consecutive accesses to each block. On an access, the stack distance is the number of other blocks between the current location of the accessed block and the top of the stack. The distance reflects the number of other blocks that were accessed between the current and the previous access to the block. A distance larger than the number of blocks that fit in each cache size represents a cache miss for that size.

In its simplest implementation, the algorithm uses a linked list to represent the stack [25]. More efficient implementations typically keep track of the reuse distances of the references instead of the references themselves, and use more sophisticated data structures, e.g. [5]. Because of the overhead of the algorithm and the need to detect all memory accesses, it can only be used on-line when hardware support is available [36], when large blocks are accessed explicitly [21], or when approximations are acceptable [34, 36]. In our caching middleware, we consider a single-level memory hierarchy (main memory), blocks (objects) that are accessed explicitly by calling the middleware, and use LRU as the object replacement policy.

6.3 Related Work

Many papers have studied the combination of dynamic workload consolidation with server turn off [8–10, 15, 31, 32]. In this paper, we demonstrate how to make energy conservation in dynamic workload consolidation more practical through the creation of a family of active low-power server states. We compare our server states against PowerNap and Somniloquy [2, 27] extensively in Sects. 6.5 and 6.6.

An orthogonal approach to consolidation and turn off is to dynamically scale the voltage/frequency of the processor (DVFS), when the CPU load is low. We focus on consolidation and turn off for two main reasons. First, DVFS currently only applies to the CPU, while other server components also consume significant power. Second, the opportunity to reduce voltage (the main source of CPU energy savings) has and will continue to diminish over time.

Our study focuses on high-performance servers with consolidated workloads requiring significant processing power. Other work has studied datacenters comprising lower performance (and power) servers [3]. These servers were not found to be particularly advantageous for Web search in terms of energy (although more advantageous in terms of cost) in [22]. More recently, Reddi et al. at [19] found that these servers perform poorly for a computationally intensive search engine workload. In Sect. 6.6.3, we compare our results to those of such servers.

Some states in the barely-alive family turn all the CPU cores off but still allow memory accesses through the network interface. Remote Direct Memory Access (RDMA) also allows memory to be accessed without host intervention. However, the previous works in RDMA have focused on using this mechanism to bypass an active CPU in fully operational servers. A few high-end network interface cards, such as InfiniBand [24], provide RDMA capabilities. Although we intentionally abstract the mechanisms required by RDMA (e.g., address registration and memory pinning) in this paper, we do rely on similar functionality.

Another approach that enables remote memory accesses in a blade chassis is disaggregated memory (DM) [23]. In DM, a set of memory blades extend the memory of the compute blades. A memory blade can be seen as a server in a barely-alive state with all cores and disks turned off. However, our approach is more flexible in that barely-alive servers can be activated and recover the full functionality of a server. In addition, most barely-alive states require no hardware modification and can use off-the-shelf clustering software. Finally, in our approach, each server includes more local memory, reducing interconnect bandwidth requirements with respect to DM.

6.4   Barely-Alive States

We propose a family of barely-alive server states. *The states differ in terms of exactly what components are turned off to conserve energy. The unifying characteristic of all states in the family is that selected levels of the memory hierarchy (main memory and possibly disks) can be accessed by remote servers, despite the fact that some components are turned off.* An Internet service can transition some servers to one of the barely-alive states, instead of the off state, after consolidating the workload on another set of servers.

6.4.1   Members of the Family

We have identified many barely-alive states, called "BA" followed by a member number. The deepest state, BA1, turns off *all* the cores, all the disks, the shared cache, all but one fan, and all but one network interface. The memory controller is kept on (even if the controller is on chip), but the memory devices are sent to the self-refresh mode immediately after any access. Remote memory accesses occur through a very low-power embedded processor built into the network interface. (Some existing network cards include programmable processors in them, e.g. [29].) This processor accesses memory by driving the memory controller directly, just as in regular DMA operations involving the network interface. In fact, compared to current server hardware, the only hardware support required by BA1 is a separate power rail for the memory controller and the low-power embedded processor (if it is not already available in the network card).

BA2 consumes slightly more power than BA1, as it manages the memory using the standard close-page policy. Under this policy, most power savings (beyond those of BA1) come from transitioning memory ranks that have no open row buffers to the (precharge) powerdown mode. BA2 requires no hardware support beyond that for BA1.

BA1 and BA2 can be used when the memory access traffic on a barely-alive server is low enough that a single network interface and embedded processor can manage. Higher load may require additional components to be activated. In state BA3, one or more additional network interfaces are activated. To name variations with different numbers of active components, we use a suffix. For example, when two active network interfaces are used, we refer to this state as BA3-2NI. Again, BA3 requires no hardware support beyond that for BA1.

If the load on a barely-alive server is excessively high for the embedded processors to handle, one or more cores (and possibly fans) must be activated; the embedded processors can be turned off. State BA4 represents these scenarios. The deepest of the BA4 states is BA4-1C, which keeps a single core, fan, and network interface active. The shared cache is active as well. In terms of hardware support, BA4 requires the ability to turn off cores independently. This ability already exists in some modern multi-core CPUs. In addition, BA4 could benefit from the ability to activate only part of the shared cache, e.g. $1/N$ of it for an N-core CPU. Current processors do not provide this feature.

One or more cores must also be active, when remote disk accesses to barely-alive servers are needed. The active core(s) can execute the device driver for the disk(s). State BA5 represents these scenarios. The deepest of the BA5 states is BA5-1C-1D which keeps a single core, fan, network interface, and disk active. BA5 requires no hardware support beyond that for BA4.

Transition overheads. The transitions to and from a barely-alive state are initiated by a CPU core (if at least one is active) or by the embedded processor (if no core is active). Transitions can be between the active state and a barely-alive state or between two barely-alive states. Regardless of the states involved, transitions can be very fast and consume little energy, since the memory contents (including any cached data and the operating system state) are not affected. In fact, updates to the memory contents can occur while the server is in a barely-alive state. The discussion below quantifies these overheads for the two extreme transitions: (1) from the active state to BA5 and back (disks remain active all the time); and (2) from the active state to BA1 and back (disks can be shut down in the barely-alive state).

The transitions between the active state and BA5 take on the order of microseconds, i.e. the time needed to transition the cores and network interfaces. The fans need not complete their transitions before the server can be declared in the barely-alive or the active state. The energy overhead of the transitions is negligible.

The transition from the active state to BA1 also takes on the order of microseconds, since the fans and disks can complete their transitions in the background. The energy overhead of this transition is dominated by the energy consumed in spinning down the disks. In contrast, the transition from BA1 to the active state is dominated in terms of both time and energy by the disk activation overheads. Carrera et al. [7] have quantified the overheads of sending an IBM Ultrastar disk to the standby state at 10 J and 2 s, and the overheads of activating it at 100 J and 10 s. Others [35] have reported much lower overheads for a Fujitsu disk. Fortunately, these overheads are modest, given that Internet service workloads allow servers to stay in a barely-alive state for long periods of time.

Implications for software. To be most useful, the barely-alive family requires the cluster software to have the ability to (1) consolidate the workload into a subset of (active) servers and (2) perform remote memory (read and/or write) accesses to barely-alive servers. For Internet services, it would be natural for the cluster software to implement some sort of cooperative main-memory caching system [6, 11, 30], which would manage the main memories of the cluster as a single large cache. This implementation could be coupled with a standard consolidation algorithm. In fact, regardless of the barely-active state(s) used, the consolidation algorithm can be the same as before [8, 31]. The only adjustment is that schemes involving larger activation overheads (e.g., on-off consolidation) require more servers to be active at all times to handle typical load spikes.

Although cooperative caching is a good application for barely-alive servers, other types of datacenter workloads are also amenable to our family of states. For example, one might implement a distributed file service that sends some servers to a barely-alive state under light load, but continues using their memories to avoid disk accesses. Another example is a replicated database system that transitions servers to a barely-alive state, but keeps updating the tables they store and/or cache. Even MapReduce computations with limited parallelism can leverage the set of main memories to store large data structures. Obviously, the best barely-alive state for these types of workloads may be different than that for cooperative caching.

When using barely-alive states in which at least one core is active (e.g., state BA4-1C), all memory addressing can be done using virtual addresses. Furthermore, the disks of barely-alive servers can be accessed (e.g., state BA5-1C-1D).

For the family members that turn off all cores (BA1, BA2, and BA3), memory addressing requires careful handling in software. In particular, as the embedded processor does not understand virtual addresses, the remote memory accesses have to specify physical addresses or be translated to physical addresses in software by the embedded processor. Memory management also becomes more difficult when multiple embedded processors are active (e.g., state BA3-2NI). In this case, the software is responsible for guaranteeing proper coordination. Finally, the embedded processor has to implement some sort of (RDMA) communication protocol to be able to receive memory access requests coming from active servers and reply to them. As our target system is a server cluster, this communication protocol can be lean and simple. Because the barely-alive states are independent of this protocol, we do not discuss it further.

6.4.2   Cooperative Caching Middleware

Our middleware implements the PRESS cooperative main-memory caching system [6], but modifies it to accommodate servers in a barely-alive state and to re-size the local caches dynamically. The goal is to reach a target cache hit ratio, while allowing energy conservation and freeing up as much memory as possible for applications. Note that the middleware cannot target an average response time, since it does not service the cache misses (as explained below). We assume that each application knows the average response time it wants to achieve, computes the target hit ratio based on this response time and the average cache hit/miss times, and informs the middleware about the computed target hit ratio.

Request distribution. The middleware caches application-level objects and names them using numerical ids. It maintains the location of each cached object in the *cooperative cache directory*, which is replicated at each server. When first received by the service, a client request is assigned to a server in round-robin fashion. This initial server decides whether to actually serve the request, depending on whether it caches the requested object. If it does not, it looks up the directory and forwards the request to a server that does (if one exists). If the remote server is in the BA2 state, the initial server accesses the object directly from its memory. If the remote server is overloaded, the initial server does not forward the request and replicates the object locally.

Applications interact with the middleware mainly by calling runtime routines for storing and fetching objects to/from the cooperative cache. A fetch call that misses the cache returns a flag reflecting the miss; in this case, the application is supposed to fetch or re-generate the object and store it in the cache. The middleware also provides calls for object invalidation. The middleware allows these calls to originate at any *active* server, i.e. servers in a barely-alive state are essentially passive "object fetch servers". The servers in the BA2 state can find objects in memory because the network interface processor shares the object addresses with the host processor. Invalidating an object cached by a barely-alive server works fine, because when the barely-alive server is activated, it realizes that the object should be invalidated by contacting one of the active nodes. To prevent the loss of cache space at the barely-alive servers in invalidate-intensive scenarios, they can be periodically activated, while some of the active servers can be sent to the barely-alive state.

Local cache re-sizing. The middleware determines the local (LRU) cache sizes that are required for a target hit ratio using the stack algorithm [25]. The middleware periodically (every hour) collects the stack information from all active servers and computes the total (cooperative) cache size required to achieve the target hit ratio. In systems that consolidate workloads and turn servers off, the middleware sets the local caches to their maximum size and informs the consolidation algorithm about the minimum number of servers (= total size divided by maximum local size) that need to remain active. In systems that use barely-alive states, the middleware sets the local cache sizes to the total cache size divided by the total number of servers.

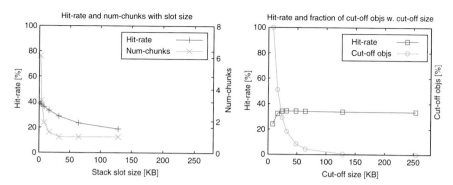

Fig. 6.1 Optimizing the caching layer

6.4.2.1 Caching Layer Optimization

The stack algorithm assumes fixed-sized cache slots (e.g. pages). In contrast, Web objects exhibit widely varying sizes. For this reason, selecting a large slot size would generate excessive fragmentation in the cache, whereas selecting a small size could generate significant overhead. Fortunately, through experimentation, we found that splitting the local sizes into two parts, each with a different slot size and associated stack, works well; splitting the cache into more parts provides only trivial benefits.

Two parameters affect the performance of this implementation, the cacheable object *cut-off size* and the cache *block-size*. The best setting for both parameters depends on the workload. The cut-off size determines the largest object that can be cached. It is important to prevent extremely large objects from being cached, because these objects tend to exhibit poor locality [6]; caching them could displace many smaller objects that exhibit better locality, thereby lowering the hit-ratio significantly. Users of our middleware configure this parameter by analyzing the object size distribution statically. Similarly, defining the best value for the slot sizes for the two parts of the local caches requires an analysis of the object size distribution, as well as the desired response time and bookkeeping overheads.

We now discuss the setting of these parameters using the snippet service as an example. Figure 6.1 (left) shows the hit-ratio (left Y axis) and average number of slot-sized chunks per object (right X axis), as a function of the cache slot size. The hit-ratio curve suggests that the best slot size would be the smallest possible. However, at this size, an average of about eight blocks are required to represent a single application object! Clearly, this number of chunks induces a significant overhead for bookkeeping. From the graph, we can see that a slot size of 8 KB translates into an average of roughly two chunks, which we use in our experiments. Figure 6.1 (right) shows the hit-ratio (left Y axis) and the number of objects which are not cached (right X axis), as a function of the cut-off size. These curves suggest that a cut-off size of 32 KB, which we use in our experimentation is the best choice for this workload.

6.4.3 Consolidation Algorithm

We use a consolidation algorithm that periodically (every hour) determines how
many servers should remain active while others can be transitioned to a low-power
state (barely-alive, Somniloquy, or off state). The behavior of the algorithm depends
on the type of low-power state the system wants to use.

For systems that use a barely-alive state, the number of active servers is based
solely on the average utilization of the resource that is closest to saturation [15, 31].
As a server-wide proxy for this average utilization, we use the average number
of outstanding requests divided by the maximum number of outstanding requests
a server can handle efficiently given the workload. Using this metric, when the
average response time increases, the utilization also increases.

When the average utilization cluster-wide is lower than the "state-transition
threshold", the algorithm tries to reduce the number of active servers. Its main
constraint is that, after consolidation, no server shall exhibit a utilization higher
than this threshold. As discussed in Sect. 6.6.1, when provisioning for potential load
spikes, the algorithm adds extra active nodes to compensate for activation delays.

For systems that use Somniloquy or off states, the number of active servers is the
maximum between the above utilization-based calculation and the hit-ratio-based
minimum number of active servers described in the previous subsection.

6.5 Qualitative Evaluation of the Barely-Alive States

Table 6.1 presents a qualitative comparison of the power consumption and transition
overheads to and from active state of the members of the barely-alive family.
The power numbers assume a single multi-core CPU and do not include power
supply losses. The table also includes the same characteristics of PowerNap [27],
Somniloquy [2], On/Off [8, 31], and low-end servers (e.g., Atom-based servers)
[3, 19, 22]. Table 6.3 shows a more detailed breakdown of the power consumptions
we assume. In comparing the systems, we assume that they run an Internet service
workload and a cluster-wide cooperative caching middleware.

We first describe the systems that rely on load consolidation (the bottom group
in the table). The barely-alive family was described in Sect. 6.2. We assume that the
content of the memory of a server with no active cores (BA1, BA2, and BA3-2NI
in the table) is only updated when the server is activated. Somniloquy is similar to
BA1, except that all accesses in the low-power state are performed to memory in the
network interface itself, rather than main memory. As a result, data updates are only
performed to main memory when the server is activated. In addition, the amount of
memory that can be accessed is limited to the size of the network interface memory.
These two characteristics mean that Somniloquy must keep more servers active than
a barely-alive system to compensate for the higher activation time and the smaller
global memory cache.

Table 6.1 State transition overheads

System	Access to all memory	Power	Transition time (up/down)	Transition energy (up/down)
Traditional servers	Y	$O(300\,\text{W})$	N/A	N/A
Low-end servers	Y	$O(50\,\text{W})$	N/A	N/A
PowerNap	Y	$O(40\,\text{W})$	$O(\mu s)/O(\mu s)$	$O(\mu J)/O(\mu J)$
BA1	Y	$O(30\,\text{W})$	$O(10\,s)/O(\mu s)$	$O(100\,\text{J})/O(10\,\text{J})$
BA2	Y	$O(40\,\text{W})$	$O(10\,s)/O(\mu s)$	$O(100\,\text{J})/O(10\,\text{J})$
BA3-2NI	Y	$O(50\,\text{W})$	$O(10\,s)/O(\mu s)$	$O(100\,\text{J})/O(10\,\text{J})$
BA4-1C	Y	$O(60\,\text{W})$	$O(10\,s)/O(\mu s)$	$O(100\,\text{J})/O(10\,\text{J})$
BA5-1C-1D	Y	$O(70\,\text{W})$	$O(\mu s)/O(\mu s)$	$O(\mu J)/O(\mu J)$
Somniloquy	N	$O(30\,\text{W})$	$O(10\,s)/O(\mu s)$	$O(100\,\text{J})/O(10\,\text{J})$
On/Off	N	$O(0\,\text{W})$	$O(100\,s)/O(\mu s)$	$O(1,000\,\text{J})/O(100\,\text{J})$

Transition overheads include the time and energy of the actual state transitions, as well as the overhead to re-load and update the memory after activation. A detailed breakdown of our assumed power consumptions is presented in Table 6.3. For a fair comparison, a disk is present in PowerNap, but never shutdown.

The On/Off system turns servers completely off after consolidation, which means that part of the cluster memory cannot be accessed, server activation takes a long time, and data updates are done in batches after activation. Thus, the On/Off system needs to keep more servers active than a barely-alive system to compensate for the smaller memory cache and guarantee that server activation does not translate into higher response times.

PowerNap and low-end servers do not rely on consolidation. PowerNap sends all components (except for disks, which were replaced by solid-state drives in [27]) to their deepest power states whenever there is any idle time at a server. Unfortunately, multi-core servers are completely idle only for very short periods of time (if at all), since the core idle times have to overlap perfectly. Low-end servers seek to provide better energy efficiency simply through the use of more efficient (and often lower performance) components; no power state changes are effected. As a basis for comparison, we also consider a system that uses traditional 1U servers and keeps them active at all times.

The key observations to make from this table are: (1) all systems have very low power states with different levels of energy savings; (2) transition overheads are not significant (except in the On/Off system), since we expect the systems that leverage consolidation to transition states at the granularity of hours. Moreover, in the barely-alive and Somniloquy systems, the time to activate a server can be reduced from $O(10\,s)$ to $O(\mu s)$, if the system does not shut down the disk; (3) when there is idle time at all (i.e., under extremely low utilization), transition frequencies are likely to be high for PowerNap, which would significantly increase the system's energy consumption; and (4) the low power of Somniloquy and On/Off is partially countered by the need to keep more servers active, leading to higher overall energy, as we shall demonstrate in our results.

Overall, the barely-alive family presents a range of interesting tradeoffs between power and overhead. BA1 is a deep power state with relatively small performance and energy overheads, whereas BA5-1C-1D consumes more power but has trivial overheads. No other system can achieve all the benefits that barely-alive states provide.

Although all of the barely-alive states support our goals well, *henceforth we will focus on BA2 only.* This state represents an interesting design point for two reasons: (1) some current processors already have a power rail for the memory controller that is separate from those for the cores; and (2) leaving one core on currently requires leaving the entire shared cache on, reducing energy savings. Both these reasons are illustrated by the Nehalem "uncore" [33].

6.6   Quantitative Evaluation of the Barely-Alive States

6.6.1   Benefits of Fast Activation

A significant challenge for all load consolidation schemes is handling typical load spikes without violating latency constraints [13, 20]. In this section, we present a simple analysis of barely-alive and previous schemes when faced with a parameterized load spike. We estimate the extra server provisioning and illustrate the tradeoffs of activation latency, standby power consumption, and data update latency. We use the intuition deriving from these tradeoffs in our detailed case study in Sect. 6.6.3.

To avoid excessive latency, extra active servers must be provisioned to absorb the spike load until more servers can be activated. The number of extra active servers must match the typical increase in load during the activation time. In more detail:

$$NumExtraAct = (MaxLoadRateAfterActTime - LoadRateBeforeSpike)/$$

$$ActServerCapacity \tag{6.1}$$

where $NumExtraAct$ is the number of extra active servers; $MaxLoadRateAfter$ $ActTime$ is the maximum request rate during a period equal to the activation time, since the beginning of a typical spike; $LoadRateBeforeSpike$ is the request rate before the spike begins; and $ActServerCapacity$ is the request processing capacity of an active server. Thus, the higher the latency of server activation, the more the request rate during the spike will increase, and the more extra servers must be provisioned.

Next, we quantify these effects. Our analysis assumes that the cluster has 32 servers, and each active server can process 2,000 connections/second and consumes 222–404 W as a linear function of utilization. For the latency of server activations, we assume 30 s for the On/Off system, and 10 s for the BA2 and Somniloquy systems. All these values match our assumptions in the simulation results section (Sect. 6.6.3).

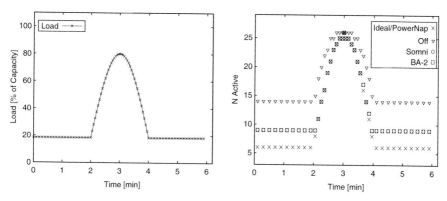

Fig. 6.2 A typical load spike (*left*) and corresponding server provisioning (*right*)

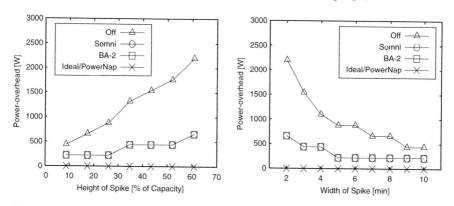

Fig. 6.3 Impact of spike height (*left*) and width (*right*) on the cluster power consumption

In Fig. 6.2 (left), we present an example of a synthetic load spike, which increases the request load on the system from 20 to 80% of its maximum capacity. Figure 6.2 (right) shows that the number of active servers before the load spike is significantly higher for the On/Off system than for the more sophisticated BA2, PowerNap, and Somniloquy systems. For a baseline comparison, the "ideal" system is an On/Off system in which servers can be brought up with zero latency and no energy overhead. As originally proposed [27], PowerNap exhibits near-zero transition latency. BA2 and Somniloquy are equivalent with respect to load spike provisioning, as long as no data needs to be modified at servers in a low-power state.

We can parameterize load spikes by duration and amplitude, and choose parameters consistent with observed behavior such as from studies of an HP customer's Web server trace [20]. Figure 6.3 (left) shows how the power overhead of extra server provisioning (with respect to the ideal system) varies with spike amplitude, assuming a duration of 2 min. We can see that the On/Off system entails a modest level of overhead with spikes of low amplitude. However, the overhead grows significantly as the spike increases in amplitude. Figure 6.3 (right) shows the impact

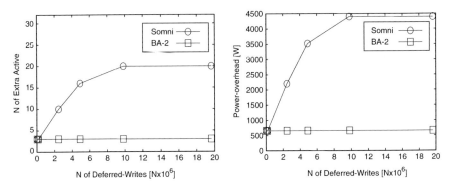

Fig. 6.4 Impact of deferred writes on the cluster provisioning (*left*) and power consumption (*right*)

of spike duration, assuming an amplitude of 60% of the peak capacity. We can see that, if the duration of the spike is 2 min, the over-provisioning overhead is large. The overhead drops to more modest levels for spikes lasting 10 min.

6.6.2 Benefits of Allowing Immediate Data Updates

Services modify data that may reside on servers that are off or in a low-power state. A key advantage of barely-alive systems is the ability to directly modify data in main memory while in a low-power state. Modification of such data is impractical in On/Off and PowerNap systems. For On/Off systems, writes would need to be source buffered and deferred to wake up, which can be problematic if systems are off for long periods of time. PowerNap can avoid this problem by waking up the server to perform data updates. However, for all but the most insignificant of write intensities, PowerNap would spend too long in active state.

Other than barely-alive, Somniloquy offers the best solution to data updates while in a low-power state. Writes can be buffered in the Somniloquy device. However, with the limited size of the Somniloquy memory (64 MB), we assume that writes would need to be buffered in their Secure Digital (SD) card auxiliary storage. The time to read the updated data from the SD card to main memory during activations increases the activation time and, thus, increases the number of extra active servers (Eq. 6.1).

In Fig. 6.4, we compare BA2 and Somniloquy as the number of deferred writes varies, assuming the same cluster and server parameters from the previous subsection. Writes are to objects typical of our Web application (6 KB each). Figure 6.4 (left) quantifies the number of extra active servers in each system. As the number of buffered writes increases (either due to higher write traffic or longer time in a low-power state), the Somniloquy activation latency becomes significant. This effect quickly results in a large number of extra active servers provisioned for spikes. The number of extra active servers levels out when transitioning servers to Somniloquy state actually starts to increase energy consumption. Figure 6.4 (right) shows the same comparison in terms of total power for extra active servers.

6.6.3 Benefits of Aggregating Memory

Although barely-alive server states provide fast server activation and allow data updates while in a low-power state, an even greater advantage is their ability to effectively use all of a cluster's memory while adjusting processing power to reduced load. In this section, we pair the systems we study with a middleware implementation of distributed cooperative object caching. The middleware manages the available memory resources across the cluster as a single large cache to avoid disk accesses. This set of experiments is motivated by the observation that, while load offered to Internet applications may vary significantly, the working set often does not.

For our evaluation, we built a trace-driven simulator and focus on a representative Internet service. We simulate the major pieces of software that are of interest, namely the service's workload, the consolidation algorithm, and the middleware for cooperative caching. We also simulate the major hardware components of the cluster (CPUs, main memory, network interfaces and switch, and disks), their utilizations, bandwidths, latencies, and power consumptions. Next, we describe these aspects of our simulator in greater detail.

6.6.3.1 Internet Service and Its Workload

Our representative application is a "snippet" generator that services Web search queries by returning a query-dependent summary of the search results. Each query generates a list of ten URLs. The snippet generator scans the pages associated with these URLs and produces a text snippet for each of them. It uses the middleware to cache the pages.

We obtained a 7-day trace representing a fraction of the query traffic directed to a popular search engine (Ask.com). Due to privacy and commercial concerns, the trace only includes information about the number of queries per second. The volume of queries follows the traditional pattern of peaks during the day and troughs during the night. Weekend traffic follows a similar pattern but with lower traffic. In order to generate a complete workload, we also analyze publicly available traces that contain all submitted queries (36.39 Million) to AOL over a duration of 3 months in 2006. The distribution of object popularity follows a Zipfian distribution [1]. We ran a sample of AOL queries against Ask.com, downloaded the content pointed to by the URLs listed in the returned results, and computed the content size. We found a median size of 6 KB following a Gamma distribution. In our experiments, we run 2 days of the trace, corresponding to a Friday and a Saturday, as well as a few extra hours of cache warm-up time pre-pended.

6.6.3.2 Discussion

We simulate a relatively simple single-tier service to demonstrate the benefits of barely-alive states and dynamic state transitions more clearly. In this service, the application data is placed in such a way that any active server can handle a cache miss by performing local disk I/O. This simplifies the consolidation algorithm and allows the system to (1) turn off the entire CPU and all disks in the barely-alive (BA2) state; or (2) transition some servers to Somniloquy or off state.

In practice, services are often more complex with multiple tiers and highly distributed data-sets. In such services, the low-power states that can be used depend on the characteristics of each tier. For example, some of the tiers may involve servers that do not require much computation or disk I/O. The system can easily transition some of those servers to deep barely-alive states (e.g., BA2). For the tiers that do require computation and disk I/O, a cache miss may cause the server to perform some local computation and communicate with other servers that will perform more computation and disk I/O. The system can still transition some of these servers to a barely-alive state (e.g., BA5-1C-1D) and perform computation and disk I/O in that state. In contrast, the other schemes would either (1) be unable to use low-power states at all; or (2) have to activate the servers first. Thus, the benefits of our server states would be even clearer for these tiers.

6.6.3.3 Simulation Methodology

We implement a detailed trace-driven simulator, which we use with our 2-day trace. We model the workload using tuples of the form *(object-id, object-size, timestamp)*. The simulator takes the workload as input and implements all aspects of the caching middleware in detail. It simulates an LRU stack and hits table for each node, which it updates using the requested *object-id*. It also simulates the memory usage accurately, using the *object-size* information. The simulator also implements the consolidation algorithm in detail.

We simulate 32-node clusters by default. We provision the clusters for the peak demand of our application. Specifically, when all servers are active, the average server utilization at the peak load intensity is roughly 70%. This setting allows enough slack to handle major unexpected increases in load. We set the default state-transition threshold to 85% of the 70% of the peak utilization, i.e. in terms of actual utilization the threshold is: $70 * 85 = 59\%$. Henceforth, when we mention a state-transition threshold, its value is always relative to the 70% peak load intensity.

The simulator models all the major hardware components of each server and the interconnect. We assume that a server's CPU utilization is directly proportional to the request load currently handled by the server. Disk utilizations and latencies are computed by accounting for average seek, rotational, and data transfer times. Similarly, the memory utilizations and latencies are computed by accounting for row buffer hits and misses. The interconnect performance is modeled by a TCP connection establishment time and its communication bandwidth.

Table 6.2 Server performance parameters

Component	Type	Performance
CPU	High-end	2.66 GHz Xeon
	Low-end	1.66 GHz Atom
Memory	DDR3	Row access: 35 ns
		Column access: 20 ns
		Row size: 2 KB
		Access size: 64 bytes
Disk	High-end	Avg seek time: 8.2 ms
		Avg rotational time: 4.2 ms
		Media transfer rate: 130 MB/s
	Low-end	Avg seek time: 11 ms
		Avg rotational time: 4.2 ms
		Media transfer rate: 155.6 MB/s
Network	Ethernet	Connection establishment:
		$(24 + 19) * 0.001$ ms
		Bandwidth: 1 Gbit/s

Table 6.3 Server power consumption breakdown by component

Component	Active	BA2	Somniloquy	Low-end
Core i7 (Xeon 5500) CPU	94–260 W	18 W (2)	2 W	
Atom (D500) CPU				0–26 W
1 Gbit/s NI	5 W	5 W	6 W (64 MB)	5 W
2 Hitachi Deskstar 7K1000	24 W	4 W	4 W	
500 MB laptop disk				10 W
DRAM	12 W (4 GB)	12 W (4 GB)	0.7 W (4 GB)	5 W (1 GB)
Fans	50 W (5)	10 W (1)	10 W (1)	10 W (1)
Small embedded CPU		1 W		
Power supply loss	37–53 W	10 W	5 W	5–8 W
	(20–15%)	(20%)	(20%)	(15%)
Total	222–404 W	60 W	28 W	35–64 W

Each simulated server has two Xeon CPUs (each with four cores), two 2 GB DIMMs of DDR3 main memory, two 7,200 rpm disks, one 1 Gbit Ethernet network interface, and five fans. We assume that the middleware is allowed to manage 1/4 of the main memory of each server. When in BA2, many of these components can be turned off. The default performance parameters of our servers are described in Table 6.2. The power consumptions of our servers in the active, BA2, and Somniloquy states are presented in Table 6.3 (servers that are off consume 0 W). These performance and power parameters came from real datasheets and papers [16–18, 22, 28]. We do not simulate PowerNap because there are very few opportunities to use it with eight cores. Note that the CPUs still consume 18 W in the BA2 state, because their memory controllers remain active. Similarly, the disks still consume 2 W each, because their interfaces need to remain on even when the disks have been spun down. We compute the power consumption of each active server as a

Table 6.4 Energy consumption and energy savings without spike provisioning

System	Weekday		Weekend-day	
	Energy (Wh/day)	Energy Savings (%)	Energy (Wh/day)	Energy Savings (%)
Baseline	229,845	0	219,108	0
BA2	169,522	26.2	144,224	34.2
On/Off	198,678	13.6	187,773	14.3
Somniloquy	218,875	4.8	216,920	1.0
Low-end	185,020	19.5	176,680	19.4

linear function of utilization between their minimum and maximum consumptions. As we can see, the power savings due to the BA2 state range from 162 to 344 W, as compared to the active state.

For further comparison, we also simulate clusters built out of lower power (and lower performance) servers that use mobile-class processors. Researchers have argued for using such servers in services, rather than workload consolidation and server turn off [3, 19, 22]. The parameters we use for these servers are listed in Table 6.3 under the "Low-end" heading.

Defining the number of low-end servers to use in a fair comparison with our system is difficult. We simulate low-end clusters six times larger than the other systems for two reasons: (1) each of the high-end servers includes two processors; and (2) previous work [19] suggested that 3× is the performance loss of low-end servers compared to single-processor high-end servers. The middleware manages 1/4 of the memory of each server (256 MB); the same ratio we use for the high-performance servers. As mentioned above, consolidation is turned off. In summary, our low-end configuration uses 192 nodes (instead of 32 nodes in the high-performance configuration) and a total cache space of 48 GB (instead of 32 GB).

6.6.3.4 Results

In our first set of simulations, we consider the case in which the systems we study are provisioned without expecting load spikes. In our second set of simulations, we consider the more realistic case in which load spikes may occur and must be provisioned for. We first identify the maximum hit ratio that can be achieved by our workload (all nodes active using their entire memories for object caching). We refer to this hit ratio as the baseline ratio. As our default, we set the target hit ratio for all systems to 95% of this baseline ratio. The total amount of cache space required by this target hit ratio is 26 GB.

Our results show that the BA2, Somniloquy, On/Off, Low-end, and baseline systems achieve an average response time within 1–2% of 23 ms during both days we study. Despite the similar performance, the energy savings achieved by these systems differ significantly. Table 6.4 lists the energy consumption and savings with

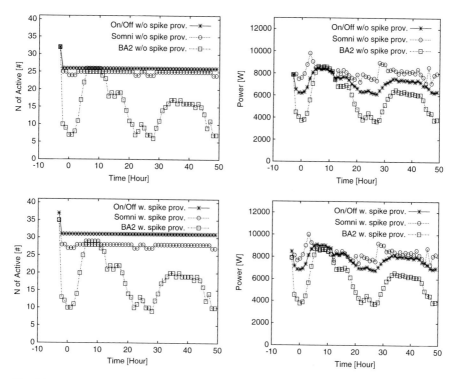

Fig. 6.5 Number of active servers (*left*) and cluster power consumption (*right*) over time, without (first row) and with (second row) spike provisioning

respect to the baseline system. As we can see from the table, in this scenario, the BA2 system achieves at least twice the energy savings of the On/Off system. The reason is that the On/Off system needs to always maintain a relatively large number of active servers to satisfy the target hit ratio. The BA2 system, on the other hand, keeps only as many active servers as necessary to service the current offered load; it transitions the other servers to the BA2 state. The top graphs in Fig. 6.5 show the number of active servers (left) and average power consumption (right) of these systems over time.

The advantage of the BA2 system is even more pronounced when we compare it against the Somniloquy system. In Somniloquy state, a server can only store 64 MB, which is only a small contribution to the global cache. For this reason, the Somniloquy system needs to keep many more active servers than the BA2 system. In fact, the former system keeps only slightly fewer active servers than the On/Off system. Again, the top graphs in Fig. 6.5 illustrate these behaviors.

The Low-end system achieves the second best energy savings; 7–15% lower savings than the BA2 system. Figure 6.5 does not show the behavior of the Low-end system because it keeps all servers active during the entire execution.

Table 6.5 Energy consumption and energy savings with spike provisioning

	Weekday		Weekend-day	
System	Energy (Wh/day)	Energy Savings (%)	Energy (Wh/day)	Energy Savings (%)
Base	229,845	0	219,108	0
BA2	171,142	25.5	145,844	33.4
On/Off	215,688	6.2	204,783	6.5
Somniloquy	224,309	2.4	222,484	−1.5

Handling load-spikes. In this set of experiments, the On/Off system is provisioned to keep some additional active nodes, so that the spikes can be handled without performance degradation. Specifically, the additional provisioning translates into five extra active nodes over time. In contrast, the BA2 system activates servers much faster so it only needs three additional active nodes. The Somniloquy system also transitions fast because the caching middleware does not perform store operations to servers that are in a low-power state. For this reason, it also only needs three extra active servers.

Table 6.5 summarizes the results assuming spike provisioning; the Baseline and Low-end systems behave as in Table 6.4, as they never transition power states and, thus, do not require extra active servers. The bottom graphs of Fig. 6.5 show the number of active nodes and power consumption over time, assuming spike provisioning. We observe that the penalty of the extra active servers hurts the On/Off system significantly more than the BA2 and Somniloquy systems. Moreover, we can see that the additional active server is enough to cause an increase in energy consumption for the Somniloquy system compared to the baseline. In contrast, the energy savings of the BA2 system decrease by only 1%.

Note that the Somniloquy results are substantially worse than in [2], where it was mainly used to keep idle desktop machines network-connected. As we discuss above, for data-intensive Internet services, Somniloquy would require much larger memory to be competitive. For services that include frequent writes, the fact that the writes would not be performed in place in Somniloquy is also a problem.

6.6.3.5 Sensitivity Analysis

In this section, we evaluate the sensitivity of our results to three key parameters: the state-transition threshold for consolidation; the ratio of active and barely-alive (BA2) powers; and the range of load intensities of the workload. Unless otherwise stated, we assume the scenario with provisioning for load spikes.

State-transition threshold. Recall that this threshold determines how much the systems consolidate; the lower the threshold, the more machines are kept active. Table 6.6 shows the results for threshold values ranging from 50 to 85%. Recall that 85% is our default threshold setting. As one would expect, the table shows that

Table 6.6 Sensitivity of the energy savings to the state-transition threshold with spike provisioning

System (transition threshold)	Weekday		Weekend-day	
	Energy (Wh/day)	Energy Savings (%)	Energy (Wh/day)	Energy Savings (%)
Base	229,845	0	219,108	0
BA2(85%)	171,142	25.5	145,844	33.4
BA2(70%)	184,926	19.5	156,823	28.4
BA2(50%)	203,984	11.3	180,025	17.8
On/Off(85%)	215,688	6.2	204,783	6.5
On/Off(70%)	221,841	3.5	204,783	6.5
On/Off(50%)	231,440	−0.7	209,371	4.4
Somniloquy(85%)	224,309	2.4	222,484	−1.5
Somniloquy(70%)	221,146	3.8	209,928	4.2
Somniloquy(50%)	222,199	3.3	201,064	8.2

the energy savings that can be achieved by the BA2 system decrease significantly, as we decrease the threshold. Nevertheless, even at the most aggressive setting (50%), the BA2 system still achieves significant energy savings (11 and 18%). The small energy savings from the On/Off system also degrade with lower thresholds. In fact, for the lowest threshold, the On/Off system actually consumes more energy than the baseline on Friday. The Somniloquy results are more interesting in that decreasing the threshold sometimes increases energy savings. The reason is that a lower threshold enables the Somniloquy system to use more memory for caching (since there are more active servers), improving its cache hit ratio.

Target hit-ratio. The target hit ratio is the main performance parameter in our systems: the higher the target, the lower the response time of the service. We consider three target hit ratios: 95 (our default), 90, and 85% of the maximum achievable hit ratio. These hit ratios require 26, 22, and 17 GB of main-memory cache in the On/Off system. For these simulations, we keep the state-transition threshold at its default value (85%).

We again find that the systems achieve average response times within a few percent of each other for each target hit ratio. In terms of energy, decreasing the target hit ratio affects the On/Off and Somniloquy systems more strongly than the BA2 system. Specifically, for a system without spike provisioning, the energy savings of the BA2 system decrease slightly from 26 to 23%, when we decrease the target hit ratio from 95 to 85%. The decrease in energy savings is a result of slightly higher server and disk utilizations. In contrast, the energy savings of the On/Off system increase from 14 to 26%, with the same change in target hit ratio. The reason for such a large improvement is that the On/Off system requires many fewer active servers with the lower target. The Somniloquy system also benefits from the lower target hit ratio, but to a smaller extent than the On/Off system. Under spike provisioning, the BA2 energy savings surpass those of its counterparts even at the 85% target hit ratio. These results illustrate that the advantage of the BA2 systems is greater when the service's performance requirements are more stringent.

Ratio of active and BA2 powers. Under our hardware assumptions, this ratio is roughly 7:1. We also studied ratios of 13:1, 3:1, and 1:1, under our default state-transition threshold. For the weekday, these ratios produce energy savings of 31.6, 15.6, and −21.8%, respectively. For the weekend day, the savings are 41.1, 20.4, and −27.9%, respectively. These results show that the BA2 system can conserve substantial energy even at a low 3:1 ratio.

Range of load intensities. Finally, we investigate the impact of the difference in load intensity between the peak and valley of the workload, assuming our default state-transition threshold and power parameters. Specifically, we scale down the difference between these load intensities by up to a factor of 4. As expected, the lower the load variation, the lower the energy savings that can be achieved. Nevertheless, the BA2 energy savings reach 10.5% on a weekday and 13.7% on a weekend day, even when the load variation is reduced by a factor of 4.

6.7   Case Study I: Mixed System

So far, we have considered systems that leverage a single low-power state for energy conservation. In this section, we propose a "mixed" system that combines off and BA2 states in the context of our cooperative caching middleware. The motivation is that the BA2 system could potentially turn servers off to conserve even more energy, when they are not needed to achieve the target hit ratio. In addition, we study the tradeoff between response time (represented by target hit ratio), state-transition threshold, and energy savings.

6.7.1   Mixed System: Off + BA2

The BA2 system we have discussed so far has an important characteristic: it minimizes the number of active servers; the number of such servers is the minimum required by the offered request load. However, the BA2 system may not require all the other nodes to be in BA2 state; it may be possible to satisfy the target hit ratio with fewer servers, and turn the others completely off. This is what our Mixed system does.

Specifically, the Mixed system activates as many servers as directed by the consolidation algorithm for a BA2 system. However, instead of using the local cache re-sizing approach of the BA2 system, it uses that of the On/Off system. In other words, instead of re-sizing the local caches of all servers so that their sum is equal to the total required cache size, it re-sizes them to their maximum size and defines the minimum number of servers that is needed to reach the total required cache size. If this minimum number is larger than the number of active servers computed by the

Fig. 6.6 Number of active
and BA2 servers in the mixed
system (without spike
provisioning)

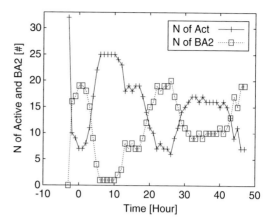

consolidation algorithm, the difference between them is the number of servers that
will be transitioned to the BA2 state. The system transitions the other servers to the
off state.

6.7.2 Results

Figure 6.6 shows the number of active and BA2 servers in our Mixed system during
the 2 days without spike provisioning. Since the target hit ratio (95% of the highest
achievable ratio) is fixed, the total number of servers in the active or BA2 state stays
constant (26) throughout the 2 days. The other servers stay in the off state.

This behavior enables the Mixed system to achieve higher energy savings than its
counterparts. In more detail, the Mixed system achieves energy savings of 30% for
the Friday and 38% for Saturday, both with respect to the baseline system. Recall
that the BA2, On/Off, and Somniloquy systems achieve energy savings of 26, 14,
and 5% for Friday, and 34, 14, and 1% for Saturday, respectively. These results
illustrate the potential energy benefits of leveraging both BA2 and off states.

We now compare the Mixed system to the BA2 and On/Off systems, as a function
of the performance level (i.e., target hit ratio) and the state-transition threshold. We
do not include the Somniloquy system because it behaves substantially worse than
the other systems. Table 6.7 presents the results broken down by day. Again, the
energy savings are computed with respect to the baseline system.

In the Mixed system, for a fixed target hit ratio, a higher state-transition threshold
enables more servers to be in the BA2 state, instead of the active state. Thus,
increasing the threshold increases the energy savings. For a fixed state-transition
threshold, a higher target hit ratio requires more servers to be in the BA2 state,
instead of the off state. Since BA2 consumes little power and the number of servers
in this situation is fairly small, the impact of varying the target hit ratio is very small
in this system. This result suggests that the Mixed system is resilient regardless of
the desired performance.

Table 6.7 Energy savings with the state-transition threshold, performance level, and day

System (transition threshold)	Weekday Energy savings (%)	Weekend-day Energy savings (%)
Target hit ratio = 95% of max		
BA2(85%)	26.2	34.2
BA2(70%)	20.2	29.2
BA2(50%)	12.0	18.6
On/Off(85%)	13.6	14.3
On/Off(70%)	10.9	14.3
On/Off(50%)	6.7	12.2
Mixed(85%)	30.3	38.4
Mixed(70%)	23.6	33.4
Mixed(50%)	14.0	22.4
Target hit ratio = 90% of max		
BA2(85%)	25	33.1
BA2(70%)	18.8	28.0
BA2(50%)	10.8	17.0
On/Off(85%)	20.3	23.8
On/Off(70%)	16.6	23.8
On/Off(50%)	9.7	16.4
Mixed(85%)	31.2	40.3
Mixed(70%)	23.8	35.2
Mixed(50%)	13.9	22.0
Target hit ratio = 85% of max		
BA2(85%)	23.4	31.8
BA2(70%)	17.4	26.9
BA2(50%)	9.7	15.5
On/Off(85%)	26.3	34.9
On/Off(70%)	20.5	30.9
On/Off(50%)	12.5	19.9
Mixed(85%)	31.4	42.4
Mixed(70%)	23.5	36.3
Mixed(50%)	13.8	21.5

We study the energy behavior of the BA2 system for a fixed target hit ratio and varying state-transition threshold in the previous section. For a fixed state-transition threshold, a higher target hit ratio increases the energy savings slightly because server the disk utilization decreases. Again, we consider the energy behavior of the On/Off system for a fixed target hit ratio and varying state-transition threshold in the previous section. For a fixed state-transition threshold, a higher target hit ratio decreases the energy savings because more servers have to stay active.

Discussion. Overall, these results demonstrate that the Mixed system consistently conserves more energy than the BA2 and On/Off systems. Compared to the BA2 system, the advantage of the Mixed system is most pronounced at low target hit ratios and high state-transition thresholds. Given these results, the Mixed system is the clear choice for services that can accept higher response times or want to conserve more energy.

Compared to the On/Off system, the advantage of the Mixed system is most pronounced at high target hit ratios and high state-transition thresholds. In this comparison, the Mixed system is the clear choice for services that require lower response times or want to conserve more energy.

6.8 Case Study II: Memory-Sharing in Barely-Alive Systems

An implicit capability of cooperative caching in Barely-Alive systems is the ability to share memory between multiple services. In this section, we examine a scenario in which two services must share the cluster. Specifically, we examine the transition when a service is running on a cluster and a second service is migrated to that cluster. This can happen in the event of a downtime at another cluster, because of maintenance or failure.

Given an allocation scheme, we can use our stack-based allocation scheme to attempt to maintain the hit-rate targets of both services. We evaluate this scenario in terms of energy savings against individual deployment of the services.

6.8.1 Memory Sharing Algorithm

Figure 6.7 shows the hit-rate as a function of our snippet service. As expected, we observe that the hit-rate is a monotonically increasing function with the memory size. Consequently, we can conclude that a linear optimization solution to maximize the aggregate application hit-rate will solve this problem optimally and in linear time

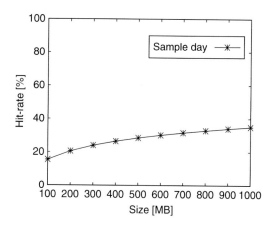

Fig. 6.7 Hit-rate curve for the snippet service

with the size of the memory stack. We therefore define the following problem: given a cluster where each node has a memory capacity M, N is the number of services to share the cluster, the hit-rate curves for each service, and a vector of size N with the minimum allowable hit-rates, we must calculate a vector of size N with the allocations for each of the services such that the aggregate hit-rate is maximized, subject to the memory capacity M.

Assume the simple case where we have the memory shared by two instances of a service which achieves 15% cumulative hit-rate with one stack slot, 20% with two stack slots and 23% with three stack slots. The minimum hit-rate is 15% for each service, which is equivalent to one stack slot, while the target hit-rate is 23% (three slots). The stack capacity is 100 slots. Initially the algorithm will run the stack algorithm for each of the applications with the 15% hit-rate and fill the allocations vector with an initial value of 1. If the initial allocations violate the stack capacity already, the algorithm fails to return valid allocations. After the initialization, the algorithm will allocate one slot at a time, to the application whose hit-rate is further away from its target hit-rate (a quantity we define as *hit-rate distance*). Round-robin is used to pick an application in the case that the hit-rate distances are equivalent.

In our example, after the initial allocations step, our algorithm will pick the first application and increase its allocation by one slot, increasing its hit-rate to 20%. In the second iteration, the second application is chosen, since its hit-rate distance to the target hit-rate is greater compared to the first application's distance (8% vs. 3%). In the third iteration, the hit-rate distances are equivalent, but because the first application was chosen first before, the algorithm now picks the second application and updates its allocation by one. Since the target hit-rate for this application is now met, the application is removed by the application list in the algorithm. The last step repeats until the target hit-rates of all applications have been met, or the memory capacity has been exhausted. Although we do not explore a priority scheme here, different priorities could be implemented by altering the round-robin ordering of allocation.

6.8.2 Results

In this section, we evaluate the energy savings when co-deploying two instances of our application on a single cluster versus each instance on its own cluster.

Figure 6.8 shows the cache allocations and hit-rates for the instances of the snippet web-services during fair memory sharing. There are two observations to make here: (1) the application hit-rates converge very fast (in less than an hour) and (2) although each application is only allocated 50% of the memory capacity, the hit-rates degrade only by very small amounts. This fact can be explained by the shape of the hit-rate curve (Fig. 6.8). The logarithmic shape of the curve guarantees that a reduction in the memory allocation yields less than proportional reduction in the hit-rate.

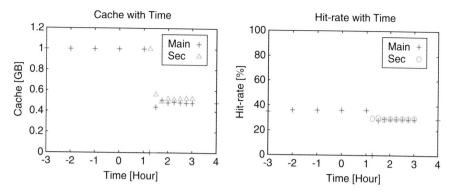

Fig. 6.8 Cache allocations (*left*) and hit-rates (*right*) for two snippet web-services during memory sharing (hours 1–4)

Fig. 6.9 Hit-rate tradeoff between two snippet web-services

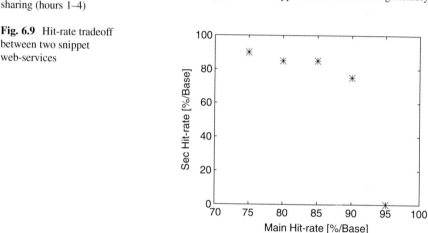

Figure 6.9 shows the hit-rate trade-off between the two applications when the main application has priority. Overall, the increase in the hit-rate of the prioritized application corresponds to a proportional decrease in the hit-rate of the other application, yet, both hit-rates are still significantly high considering the reductions in their memory allocations, confirming the scalability of our previous observation. At some point, however, the target hit-rate of the prioritized application is so high that the second application cannot satisfy its minimum target hit-rate. In our graph, this is when the main application targets a hit-rate of 95%, which corresponds to a memory allocation of 0.8 GB, while the minimum target hit-rate is 85% and requires an allocation of 0.5 GB.

In terms of energy consumption, the barely-alive sharing cluster consumes energy close to the barely-alive system with the consolidation threshold set to 50%, which is 207,461 KWh for the weekday and 185,119 KWh for the weekend-day. Had we deployed each of the web-services on two separate mixed clusters, and had we picked the optimal configuration for each of those clusters, the respective energy consumptions would have been $2 * 157,693 = 315,386$ KWh and $2 * 126,223 = 252,446$ KWh. Therefore, the barely-alive sharing system saves 34.22% energy over a weekday and 26.78% over the weekend-day, compared to the optimal, individual cluster deployment.

6.9 Conclusion

In this paper, we introduced the barely-alive family of low-power server states. We compared the family to conventional on-off consolidation, other low-power server schemes, and low-end servers. We found that the ability to access memory while in a low-power state has important advantages for both keeping data current and for cooperative caching. Our study of an Internet service workload with cooperative caching showed that conserving energy by using only a barely-alive state can save significant energy, up to 34%. Energy savings can be even higher, up to 38%, when the service may transition servers to either a barely-alive or the off states.

References

1. Adamic L (2000) Zipf, power-laws, and Pareto – a ranking tutorial. Technical report, HP Labs
2. Agarwal Y et al (2009) Somniloquy: augmenting network interfaces to reduce pc energy usage. In: Proceedings of the 6th USENIX symposium on networked systems design and implementation, Boston. USENIX Association, Berkeley, pp 365–380. http://dl.acm.org/citation.cfm?id=1558977.1559002
3. Andersen D et al (2009) FAWN: a fast array of wimpy nodes. In: Proceedings of the ACM SIGOPS 22nd symposium on operating systems principles, SOSP'09, Big Sky. ACM, New York, pp 1–14. doi:http://doi.acm.org/10.1145/1629575.1629577. http://doi.acm.org/10.1145/1629575.1629577
4. Barroso LA, Hölzle U (2007) The case for energy-proportional computing. Computer 40:33–37. doi:10.1109/MC.2007.443. http://dl.acm.org/citation.cfm?id=1339817.1339894
5. Bennett BT, Kruskal VJ (1975) LRU stack processing. IBM Res J 19(4):353–357
6. Carrera EV, Bianchini R (2005) PRESS: a clustered server based on user-level communication. IEEE Trans Parallel Distrib Syst 16:385–395. doi:http://dx.doi.org/10.1109/TPDS.2005.60. http://dx.doi.org/10.1109/TPDS.2005.60
7. Carrera EV et al (2003) Conserving disk energy in network servers. In: Proceedings of the 17th annual international conference on supercomputing, ICS'03, San Francisico. ACM, New York, pp 86–97. doi:http://doi.acm.org/10.1145/782814.782829. http://doi.acm.org/10.1145/782814.782829
8. Chase J et al (2001) Managing energy and server resources in hosting centers. In: Proceedings of the eighteenth ACM symposium on operating systems principles, SOSP'01, Banff. ACM, New York, pp 103–116. doi:http://doi.acm.org/10.1145/502034.502045. http://doi.acm.org/10.1145/502034.502045
9. Chen Y et al (2005) Managing server energy and operational costs in hosting centers. In: Proceedings of the 2005 ACM SIGMETRICS international conference on measurement and modeling of computer systems, SIGMETRICS'05, Banff. ACM, New York, pp 303–314. doi:http://doi.acm.org/10.1145/1064212.1064253. http://doi.acm.org/10.1145/1064212.1064253

10. Chen G et al (2008) Energy-aware server provisioning and load dispatching for connection-intensive internet services. In: Proceedings of the 5th USENIX symposium on networked systems design and implementation, NSDI'08, San Francisco. USENIX Association, Berkeley, pp 337–350. http://dl.acm.org/citation.cfm?id=1387589.1387613
11. Dormando: Memcached (2011). Http://memcached.org
12. Elnozahy E et al (2003) Energy-efficient server clusters. In: Proceedings of the 2nd workshop on power-aware computer systems, PACS'02, Cambridge. Springer, Berlin/Heidelberg, pp 179–197. http://dl.acm.org/citation.cfm?id=1766991.1767007
13. Fan X et al (2007) Power provisioning for a warehouse-sized computer. In: Proceedings of the 34th annual international symposium on computer architecture, ISCA'07, San Diego. ACM, New York, pp 13–23. doi:http://doi.acm.org/10.1145/1250662.1250665. http://doi.acm.org/10.1145/1250662.1250665
14. Fitzpatrick B (2004) Distributed caching with memcached. Linux J 2004:5. http://dl.acm.org/citation.cfm?id=1012889.1012894
15. Heath T et al (2005) Energy conservation in heterogeneous server clusters. In: Proceedings of the 10th ACM SIGPLAN symposium on principles and practice of parallel programming, PPoPP'05, Chicago. ACM, New York, pp 186–195. doi:http://doi.acm.org/10.1145/1065944.1065969. http://doi.acm.org/10.1145/1065944.1065969
16. Hitachi: Deskstar 7k1000 (2011). Specification Sheet
17. Intel: Intel Xeon processor 5500 series datasheet, volume 1 (2009)
18. Intel: Intel Atom processor d400 and d500 series datasheet, volume 1 (2010)
19. Janapa Reddi V et al (2010) Web search using mobile cores: quantifying and mitigating the price of efficiency. In: Proceedings of the 37th annual international symposium on computer architecture, ISCA'10, Saint-Malo. ACM, New York, pp 314–325. doi:http://doi.acm.org/10.1145/1815961.1816002. http://doi.acm.org/10.1145/1815961.1816002
20. Jung G et al (2009) A cost-sensitive adaptation engine for server consolidation of multitier applications. In: Proceedings of the ACM/IFIP/USENIX 10th international conference on middleware, Middleware'09, Urbana. Springer, Berlin/Heidelberg, pp 163–183. http://dl.acm.org/citation.cfm?id=1813355.1813367
21. Kim JM et al (2000) A low-overhead high-performance unified buffer management scheme that exploits sequential and looping references. In: Proceedings of the 4th conference on symposium on operating system design & implementation – volume 4, OSDI'00, San Diego. USENIX Association, Berkeley, pp 9–9. http://dl.acm.org/citation.cfm?id=1251229.1251238
22. Lim K et al (2008) Understanding and designing new server architectures for emerging warehouse-computing environments. In: Proceedings of the 35th annual international symposium on computer architecture, ISCA'08, Beijing. IEEE Computer Society, Washington, DC, pp 315–326. doi:http://dx.doi.org/10.1109/ISCA.2008.37. http://dx.doi.org/10.1109/ISCA.2008.37
23. Lim K et al (2009) Disaggregated memory for expansion and sharing in blade servers. In: Proceedings of the 36th annual international symposium on computer architecture, ISCA'09, Austin. ACM, New York, pp 267–278. doi:http://doi.acm.org/10.1145/1555754.1555789. http://doi.acm.org/10.1145/1555754.1555789
24. Liu J et al (2003) High performance RDMA-based MPI implementation over infiniBand. In: Proceedings of the 17th annual international conference on supercomputing, ICS'03, San Francisico. ACM, New York, pp 295–304. doi:http://doi.acm.org/10.1145/782814.782855. http://doi.acm.org/10.1145/782814.782855
25. Mattson RL et al (1970) Evaluation techniques for storage hierarchies. IBM Syst J 9:78–117. doi:http://dx.doi.org/10.1147/sj.92.0078. http://dx.doi.org/10.1147/sj.92.0078
26. Meisner D, Wenisch TF (2012) Dreamweaver: architectural support for deep sleep. In: Proceedings of the seventeenth international conference on architectural support for programming languages and operating systems, ASPLOS'12, London. ACM, New York, pp 313–324. doi:10.1145/2150976.2151009. http://doi.acm.org/10.1145/2150976.2151009

27. Meisner D et al (2009) PowerNap: eliminating server idle power. In: Proceeding of the 14th international conference on architectural support for programming languages and operating systems, ASPLOS'09, Washington, DC. ACM, New York, pp 205–216. doi:http://doi.acm.org/10.1145/1508244.1508269. http://doi.acm.org/10.1145/1508244.1508269
28. Micron: System power calculators (2011). http://www.micron.com/support/dram/power_calc.html
29. Myricom: Myrinet (2009). http://www.myri.com/myrinet
30. Pai V et al (1998) Locality-aware request distribution in cluster-based network servers. In: Proceedings of the eighth international conference on architectural support for programming languages and operating systems, ASPLOS-VIII, San Jose. ACM, New York, pp 205–216. doi:http://doi.acm.org/10.1145/291069.291048. http://doi.acm.org/10.1145/291069.291048
31. Pinheiro E et al (2001) Load balancing and unbalancing for power and performance in cluster-based systems. In: Proceedings of the workshop on compilers and operating systems for low power, COLP'01, Barcelona
32. Rajamani K, Lefurgy C (2003) On evaluating request-distribution schemes for saving energy in server clusters. In: Proceedings of the 2003 IEEE international symposium on performance analysis of systems and software, Austin. IEEE Computer Society, Washington, DC, pp 111–122. http://dl.acm.org/citation.cfm?id=1153924.1154555
33. Shimpi, A.L.: Nehalem: the unwritten chapters. AnandTech (2008)
34. Tam DK et al (2009) RapidMRC: approximating L2 miss rate curves on commodity systems for online optimizations. In: Proceeding of the 14th international conference on architectural support for programming languages and operating systems, ASPLOS'09, Washington, DC. ACM, New York, pp 121–132. doi:http://doi.acm.org/10.1145/1508244.1508259. http://doi.acm.org/10.1145/1508244.1508259
35. Weddle C et al (2007) PARAID: a gear-shifting power-aware RAID. ACM Trans Storage 3. doi:http://doi.acm.org/10.1145/1289720.1289721. http://doi.acm.org/10.1145/1289720.1289721
36. Zhou P et al (2004) Dynamic tracking of page miss ratio curve for memory management. In: Proceedings of the 11th international conference on architectural support for programming languages and operating systems, ASPLOS-XI, Boston. ACM, New York, pp 177–188. doi:http://doi.acm.org/10.1145/1024393.1024415. http://doi.acm.org/10.1145/1024393.1024415

Chapter 7
Energy Storage System Design
for Green-Energy Cyber Physical Systems

Jie Wu, James Williamson, and Li Shang

7.1 Introduction

Electric-drive transportation offers a wonderful new opportunity [1,2] to address air-pollution issues and petroleum consumption problems around the world. Currently, the greenhouse gas emissions from conventional transportation account for 40% of air-pollution emissions from all energy-using sectors [3, 4]. Development of new electric-drive techniques, in the transportation sector, is both a new and ongoing endeavor. Hybrid electric vehicles (HEVs) have been quickly adopted and widely deployed over the past decade. Presently, plug-in hybrid electric vehicles (PHEVs), which use the electricity from the electric power grid along with petroleum to power the vehicle, have received considerable recent attention to significantly reduce petroleum consumption and greenhouse gas emissions.

Since fundamental challenges are raised in (P)HEV development—including how to store the electric energy from the electric power grid, how to power the vehicle in an electric mode without using the combustion engine, or how to run in a combination mode—the electric energy storage system (ESS) has become a key component for fuel displacement potential in (P)HEV design [1,5,6]. However, due to the imbalance between the fast-growing energy demand and the ESS supply, energy storage technology has been the key bottleneck in (P)HEV design. More specifically, the ESS high cost, limited energy capacity [7], limited life time [8,9], and safety are major obstacles that need overcoming for (P)HEV to have market penetration. Therefore, this chapter discusses the challenges of energy storage systems in the (P)HEV application area.

J. Wu (✉) • J. Williamson • L. Shang
Department of Electrical, Computer, and Energy Engineering,
University of Colorado Boulder, Boulder, CO 80309, USA
e-mail: Jie.Wu-2@Colorado.EDU; james.a.williamson@colorado.edu; li.shang@colorado.edu

P.P. Pande et al. (eds.), *Design Technologies for Green and Sustainable Computing Systems*, 179
DOI 10.1007/978-1-4614-4975-1_7, © Springer Science+Business Media New York 2013

7.2   Motivation and Rationale

This section overviews the electric energy storage technologies for (P)HEV design
and summarizes the challenges of ESS.

7.2.1   Energy Storage Technologies Overview
and Performance Metrics

Energy storage Systems (ESS) are significant components in electric vehicles (EVs),
HEVs, and PHEVs [10–12]. In ESS-aware (P)HEVs, the power density and energy
density of the ESS needs to be high enough to satisfy the power and energy demand.
However, no single energy storage element can fulfill the energy requirement of
(P)HEVs. A typical way to address this issue is to build a large scale ESS with
over 1,000 energy storage elements, connected in parallel and series, to supply high
power and energy. This system is controlled by a distributed energy management
system [13], which measures and monitors the run-time current, the run-time voltage
and the temperature of the energy storage element. The following performance
metrics are considered in large-scale ESS modeling, optimization, and control for
green-energy electric-drive vehicles [14].

- **Cost:**
  The primary obstacle to market penetration of (P)HEVs is the cost of large-scale
  ESS. The cost goals selected for the ESS is set to be substantially challenging
  to promote (P)HEV technology development. For instance, high-energy batteries
  cost from \$800/kWh to \$1,000/kWh [15]. The price of ESS is more than half
  price of (P)HEV. The cost of large-scale ESS is an essentially important concern
  when modeling, optimizing, and controlling ESS. This is because the ESS
  cost is strongly correlated with its run-time charge-cycle efficiency and long-
  term lifetime. The former metric determines run-time fuel savings; the latter
  assesses the ESS overall lifetime and financial return. Moreover, the cost of
  ESS is also contributed by ESS configuration, such as the number of storage
  units and the type and size for each energy storage unit. That means different
  ESS configurations generally lead to different ESS costs. The overall ESS cost
  must be minimized by appropriate ESS configuration and while satisfying some
  important constraints, such as total energy storage capacity, peak power demand,
  long-term lifetime, run-time charge-cycle efficiency, and safety.
- **Energy density:**
  Energy density determines the maximal available energy within an ESS charge
  cycle. Among the electrochemical energy storage technologies, Li-ion batteries
  have the highest energy storage densities. The specific energy density of Li-
  ion batteries is typically around 200 Wh/kg, while ultracapacitors and NiCd/
  NiMH batteries have lower energy densities of <50 Wh/kg and 60–80 Wh/kg,

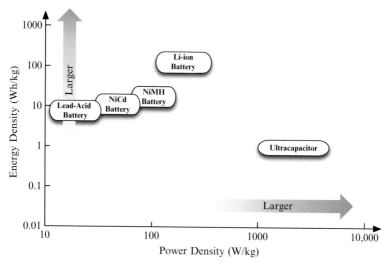

Fig. 7.1 The power and energy density for different energy storage technologies

respectively [16]. Therefore, due to their high energy densities, Li-ion batteries are proving to be the most advantageous energy storage technology for (P)HEVs.

- **Peak power:**
  Peak power is calculated as the maximal instantaneous power that can be delivered by a large-scale ESS to a vehicle. Note that energy density and peak power are two distinct performance metrics. In comparison to the Li-ion battery, the ultracapacitor provides exceptional power density due to the charges being physically stored on the electrodes [5, 16], as shown in Fig. 7.1. Therefore, ultracapacitors can be seen in (P)HEV applications with high peak or pulse power requirements. Since large-scale ESS designs must leverage the high energy density and high power density of energy storage units to respectively support stringent energy capacity requirements and high run-time peak power demand, a list of energy and power densities of various storage energy units may act as a guideline when choosing an appropriate ESS configuration.

- **Run-time charge-cycle efficiency:**
  Run-time charge-cycle efficiency is defined as the ratio of the run-time charged/discharged energy capacity to the rated energy capacity of an ESS over the entire charging-discharging process. The run-time charge-cycle efficiency is directly affected by run-time usage and the ambient environment, which is in turn determined by the user-centric run-time driving profiles, e.g., speed, acceleration, slope, and trip duration. For instance, aggressive driving profiles lead to frequent charging or discharging of the ESS, which results in dramatic ESS temperature increases as well as deterioration of its potential energy delivering capability, thus decreasing the ESS lifetime and ESS charge-cycle efficiency [17–19]. Therefore, accurate analysis of the ESS must be driven by real-world user driving studies.

– **Lifetime:**

ESS lifetime is characterized as long-term cycle life, i.e., the total number of charge-discharge cycles before the system capacity permanently degrades below a certain threshold. It is desirable that the ESS has a slow deterioration of energy capacity. This deterioration occurs during the charging-discharging process when unavoidable and unwanted chemical reactions carry out inside the energy storage unit. ESS lifetime is mainly affected by temperature and depth-of-discharge, which characterize the ESS aging effects. The most dominant aging effects have strong correlation with temperature [20]. Since large-scale ESS is comprised by a large number of energy storage units, the system capacity and lifetime are determined by the weakest of those units. Variances determined by manufacture tolerance and the heterogeneity of the environment lead to significant energy capacity degradations and variations among individual energy storage units. Without the careful balancing of those energy storage units, an individual unit can overcharge or over discharge over time. The total energy capacity then decreases rapidly during operation, which could result in the failure of the ESS. However, (P)HEVs impose stringent lifetime constraints on the ESS. For instance, automotive vendors typically target a 10~20-year ESS lifetime guarantee with 20~30% maximal capacity degradation [21]. Due to irreversible physical and chemical changes in Li-ion battery technology, the aging effects of Li-ion batteries are the primary lifetime concern in ESS design and optimization. Different from the Li-ion battery technology, the ultracapacitor demonstrates excellent lifetime [22] due to the fact that there are no chemical changes on the electrodes.

7.2.2 Challenges of ESS Design

ESS modeling, optimization, and control design is necessary to combine all discussed performance metrics together. More specifically,

– Considering that the Lithium-ion battery is the current fundamental energy storage technology, high ESS lifetime is a crucial challenge in ESS modeling, optimization, and control design. A key issue concerning ESS lifetime is run-time aging effects, which are exponential functions of temperature [19, 23]. ESS temperature dramatically increases under frequent high-rate charge and discharge operations [12, 24]. Therefore, to improve ESS lifetime, effectively handling run-time surge currents becomes vital to consider. An effective way to address this issue might be to increase the number of units in the ESS, which would mitigate the high-rate charging–discharging current. However, such a strategy would also increase ESS cost.

– The performance of the ESS (e.g., its energy capacity), the run-time charge-cycle efficiency, and the lifetime are all limited by its weakest element. Without balancing performance metrics across the entire system, such as temperature gradients across the pack, mismatched degradation over calendar life, and manufacturing tolerance, individual units can tend toward overcharge and excessive discharge [25, 26], resulting in severe lifetime limits and potential failure of the overall system. This scenario becomes worse as the number of ESS elements increases.

- Moreover, different motorists provide variances in (P)HEV operation, resulting in substantially different ESS run-time use and aging effects. For instance, intensive run-time use of the ESS by aggressive driving causes considerable self-heating in the ESS, which accelerates the temperature-dependent aging effects and ultimately leads to ESS energy capacity degradation. Therefore, data taken from real-world motorists helps to quantify (P)HEV energy usage. This information can also help to give vehicle companies better strategies in building the most effective ESS for the (P)HEV.

In short, the large-scale ESS has emerged as a fundamental factor in the (P)HEV. To meet the stringent energy budget of these vehicles, researchers have explored various ESS modeling, architecture design, system-level optimization, and control design to effectively power (P)HEVs.

The rest of this chapter is organized as follows. Section 7.3 presents the ESS modeling overview of current research. Section 7.4 describes the ESS-Aware motorist driving analysis in the (P)HEV. Section 7.5 discusses the ESS architecture design framework. Section 7.6 presents the recent studies in system-level ESS optimization. The chapter concludes in Sect. 7.7.

7.3 ESS Modeling Overview

This section presents a large-scale ESS modeling overview found in recent studies. Specifically, this section is organized as follows: Sect. 7.3.1 introduces the ESS unit's major effects. Section 7.3.2 describes basic large-scale system-level ESS modeling.

7.3.1 ESS Unit Major Effects

This subsection overviews the major effects of the ESS unit. In view of the sophisticated electrochemical characteristics of the ESS, the major electrochemical properties for each energy storage unit [27–32] include run-time rate capacity effects, run-time recovery effects, self-discharge effects, and capacity fading effects, as described as follows.

Because the energy drawn from a battery is not always equivalent to the energy consumed in device circuits, understanding discharge behavior is essential for optimal system design.

- **Run-time Rate Capacity Effect:**
  Because battery lifetime shortens as discharge rate increases, run-time discharge rate becomes an essential component of ESS design. During the discharge procedure, a fully charged energy storage unit with the maximum concentration of active reaction sites in the cathode is connected to a load and the current flows through the external circuit. If the current discharge profile (i.e., rate of discharge) is high, the reduction occurs only at the outer surface of the cathode. Thus, this results in the lower concentration of active species at the electrode surface. When the concentration falls below a certain threshold, which corresponds to the voltage cutoff, the electrochemical reaction can no longer be sustained at the cathode. The potential energy delivering capability is deteriorated and battery lifetime decreases. This effect, named *rate capacity effect* [20], represents the dependency between the actual capacity of a battery and the magnitude of its current discharge profile [18]. Due to its structure and characteristics, run-time discharge rate has little impact on the ultracapacitor [33].

- **Run-time recovery effect:**
  In the ESS, the run-time charging–discharging procedure is a burst process, followed by an idle period. During this time slot, the unused charge in an electrode will be available when the concentration gradient flattens out after enough time. Therefore, based on the existing studies [18, 29, 34, 35], the ESS unit partially recovers its capacity lost during the run-time idle periods [17]. This phenomenon, named *recovery effect*, occurs when the reactants have sufficient time to diffuse into the electrode surface and leads to more ions in the idle time.

- **Self-Discharge Effect:**
  The self-discharge effect of the energy storage unit is defined as the gradual decrease in voltage that occurs when the energy storage unit is *unconnected* to either a charging circuit or an electrical load [36, 37]. It is an important factor in determining the duration of maintaining stored energy for *low-duty-cycle applications* [38], as opposed to run-time energy usage.

  The self-discharge effect for the Li-ion battery is generally ignored due to its negligible impact [33]. Meanwhile, this effect has been studied in the ultracapacitor to determine the loss of stored energy [36–38] and the decrease in voltage, modeled as a non-linear function of time due to leakage charges.

  When the ultracapacitor is in use, the energy consumption of run-time usage dominates the energy leakage, as such becoming much more important to analyze and optimize. Therefore, little work in the literature considers the self-discharge effect while the ultracapacitor is connected to the circuit.

- **Capacity Fading Effect:**
  In contrast to the long lifetime with little degradation over millions of charge cycles [22] of electric double layer capacitors (i.e., ultracapacitors), the capacity fading effect of the Li-ion battery is a major lifetime concern. The capacity fading effect is defined as the effect of battery capacity losses to a specific percentage of its initial fully charged capacity. The long-term lifetime of the Li-ion battery is affected by the capacity fading effect, such as electrolyte decomposition and cell oxidation [20, 23, 39]. Based on the properties of the Li-ion battery, the

oxidation procedure is the most important effect, resulting in a solid-electrolyte interphase (SEI) film in the electrode surface. As this passive film grows thicker over time, the internal resistance of the Li-ion becomes larger and leads to worse battery long-term capacity. Obviously, this phenomena of the ESS unit is closely correlated with the replacement cost of the ESS. Note that the long-term lifetime of a Li-ion battery depends on the temperature and depth-of-discharge (DOD) [23,32,39,40]. The former metric determines the chemical activity which impacts on the internal resistance, reducing the capacity of the Li-ion battery; the latter influencing open-circuit voltage and internal resistance. These irreversible scenarios ultimately cause whole system failure.

7.3.2 ESS Model Design Overview

Recently, several researchers have focused on the modeling of an ESS to characterize ESS run-time charging–discharging behaviors and long-term performance. This chapter categories those models to two basic criteria:

- **Single ESS modeling in portable embedded systems:**
  Since the 1990s, several works have developed models for portable embedded systems with small size, such as mobile devices. This chapter introduces single ESS modeling in detail, including physical-level models, system-level analytical models, empirical models, equivalent circuit models, and stochastic models.
- **Large-scale ESS modeling in (P)HEV applications:**
  From the viewpoint of the (P)HEV application, large-scale ESS modeling design has been given new consideration in current years. In comparison to the portable embedded system, the (P)HEV involves a large number of energy storage units (i.e., over 1,000) to power vehicles with the necessary high output voltage, current, and energy storage capacity. Therefore, direct extensions of existing single ESS models to the (P)HEV application may not satisfy the requirement of computationally feasible mathematics in capturing the interaction effect among those units. Therefore, this chapter describes the large-scale ESS modeling design for the (P)HEV application, including the system-level large-scale ESS model and the circuit-level large-scale ESS model.

7.3.2.1 Single ESS Unit Model Overview

This subsection introduces the existing work which have focused on single ESS unit modeling.

1. **Physical-level model:**
   Physical-level models present the detailed physical procedures occurring in the ESS. It is the most accurate model to characterize the physical parameters of the ESS. Due to a complex and slow capturing process, the physical-level models

are computationally prohibitive for ESS designers. Even for simple structures, a purely physical model is well beyond any reasonable implementation complexity.

Dualfoil et al. [41] developed a Fortran program to calculate the short-term lifetime, run-time potential degradation, and concentration variation of the Li-ion battery from the electrochemical level. This program supplies a simulation tool to accurately capture the major physical factors for single ESS unit designers.

Vetter et al. [20] conducted a detailed physical analysis of the Lithium-ion battery's electrochemical capacity fading phenomenon. This model describes the capacity degradation mechanism from concentration solution theory under different storage scenarios based on the positive active materials in the cathode. This work shows that capacity fading is highly dependent on cycling, temperature, and storage conditions.

Smith et al. [42] designed a physical model, using a linear kalman filter scheme to reduce the existing high order electrochemical model so as to obtain the concentration gradients, internal potential drops, and state-of-charge. The authors used the model order reduction method to develop a seventh-order Li-ion battery model to fast solve the physical governing equations of Doyle's physical model [27, 43].

2. **System-level analytical model:**

   System-level analytical models capture system-level behavior of Lithium-ion batteries during the charging or discharging procedure, and do so from a macro perspective. Although the accuracy of these types of models is less than the physical-level model, the computational complexity is better. Physical simulation of a given Li-ion battery requires estimation of more than 50 physical parameters, even when given explicit knowledge of some Li-ion specific information including chemical composition, capacity, battery structure, temperature, and other characteristics.

   Rakhamatov et al. [30, 44] presented an analytical model for lifetime prediction of battery cells under time-varying load conditions. This analytical model considered that battery lifetime is closely related to time-varying load, resulting in concentration of active reactions changes. Its changing process consists of two main procedures: electrochemical reaction at the electrode surface and diffusion in the electrolyte. The authors used the proposed analytical model to describe reaction behavior and to build system-level diffusion equations in order to characterize the diffusion process.

   Peng et al. [23] proposed an analytical model that uses battery charge/discharge history information to estimate remaining capacity, including temperature effect and Li-ion aging effect under variable workload conditions. This model used temperature-dependence kinetics theory to capture the capacity fading behavior of the Li-ion battery, taking into account temperature-dependence SEI film growing, number of cycles, and discharge rate effects.

3. **Empirical model:**

   Based on experimental data from physical measurements of ESS, empirical models can easily build and fast predict run-time charging or discharging behaviors of ESS. However, due to these models being based on observation data and without deep theoretical support, the accuracy is lower than physical and analytical models. While this is the general case, the empirical model does work well in some specific scenarios.

Gao et al. [45] proposed a Li-ion battery model from experimental data with low temperatures and high discharge rates, taking into account nonlinear equilibrium potentials, temperature-dependence of Li-ion capacities, thermal effects and response to transient power demands. This presented model adopted empirical equations to capture the behavior of the thermal effect of Li-ion batteries.

Santhanagopalan et al. developed two empirical electrochemical models to characterize the Lithium-ion battery cell aging process occurring during battery charge–discharge cycles [39]. Incorporation of the parabolic approximation for the solid phase concentration of the diffusing species significantly reduces the computational time as compared to the physical model developed by Fuller et al. [27].

4. **Equivalent circuit model:**

In this sub item, this chapter considers a class of techniques that model battery charging–discharging behavior using an equivalent electrical circuit, instead of empirical approximation or describing the electrochemical processes of the energy storage unit. The equivalent circuit model attempts to provide an equivalent representation of an energy storage unit. Despite acceptable accuracy and computational complexity, these models have limited utility for automated design space exploration because they lack analytical expressions for many variables of interest.

The first electrical-circuit models were proposed by Hageman [46]. He used simple PSpice circuits to simulate nickel-cadmium, lead-acid and alkaline batteries. The basic idea of those battery models is that a capacitor represents the capacity of the battery and a discharge-rate normalizer determines the lost capacity at high discharge currents.

Liaw et al. [47] proposed an equivalent electric circuit model to simulate charge and discharge behaviors of the lithium-ion battery. This model characterized the SOC-dependent open-circuit voltage and resistance with an equivalent circuit to obtain accurate prediction of unit discharge behavior.

Lee et al. [48] formulated ways to extract the state-of-charge (SOC) factor by an equivalent electric circuit model. This model built a bridge between the open-circuit voltage and the state-of-charge to accurately estimate the capacity of a Lithium-ion battery.

5. **Stochastic model:**

A stochastic model of a battery is described in [18], where the battery is represented by a finite number of charge units, and the discharge behavior of the battery is modeled using a discrete-time transient stochastic process.

The first stochastic battery models were developed by Chiasserini et al.. Between 1999 and 2001, they published a series of papers on battery modeling based on discrete-time Markov chains [49–52]. In these papers, the stochastic battery models were applied in a portable mobile communication device.

Rao et al. [29] proposed a stochastic battery model based on the analytical Kinetic Battery Model (KiBaM) proposed by Manwell et al. [53]. This model captured the battery behavior as a Markov process with probabilities in terms of physical-based parameters of the energy storage unit. The parameters used for this model are determined by a pretest, which takes into account the newfound background into recovery and rate capacity effect.

7.3.2.2 Large-Scale ESS Model Overview

From the viewpoint of (P)HEV application, the modeling of large-scale ESS has drawn significant attention in recent years. This chapter introduces the large-scale ESS modeling in detail, including the equivalent circuit model and analytical model.

1. **Equivalent circuit model:**
   National Renewable Energy Laboratory (NREL) [28] proposed an equivalent circuit model to predict the current, voltage, SOC, and temperature of a battery. This model considered many parameters, including temperature variation, voltage limits, and an SOC calculator. It is accurate to characterize the charging/discharging cycles of a battery in actual driving cycles by ADVISOR tools.

   Argonne National Laboratory [54] developed an equivalent electric circuit model for Lithium-Manganese Spinel/Lithium-Titanate batteries to capture their run-time charging–discharging behaviors. They used the powertrain system analysis tool (PSAT) [55] with MATLAB/Simulink to estimate the run-time performance of batteries.

   Kroeze et al. [56] designed a battery model using the equivalent circuit method to capture self-discharge and run-time charging behaviors. The proposed model represented major lithium-ion battery run-time behaviors within a dynamic hybrid electric vehicle (HEV) simulator.

2. **System-level analytical model:**
   The system-level analytical model distinguishes itself from the existing equivalent electric circuit model, not only because it builds the large-scale model of battery from the top-level based on physical and mathematical theory, but also because it models several effects that are important to the large-scale ESS cost and lifetime.

   Wu et al. [19, 57] developed a large-scale system-level analytical ESS model to characterize ESS long-term capacity fading effects, their impacts on ESS long-term cycle life, and interaction effects, such as the variances in manufacture tolerance and the environment's heterogeneity. This study developed the ESS short-term quasi-static model, which consists of the run-time thermal model and run-time charge capacity and aging model, to characterize the ESS run-time performance and thermal effect. Given the run-time current charge-discharge profile and the configuration of the ESS, this study has proposed the ESS run-time thermal model as follows:

$$\mathbf{C}_{heat[N\times N]} \cdot \frac{d\mathbf{T}_{[N\times 1]}(t)}{dt} = \mathbf{G}_{[N\times N]} \cdot \mathbf{T}_{[N\times 1]}(t) + \mathbf{P}_{[N\times 1]}U(t), \qquad (7.1)$$

where matrix $\mathbf{C}_{heat[N\times N]}$ models the heat capacity of the N units. Matrix $\mathbf{G}_{[N\times N]}$ models the thermal conductance between adjacent units. $\mathbf{P}_{[N\times 1]}$ models the run-time power dissipation of individual units. $U(t)$ is a step function. The proposed thermal model characterizes the heterogeneous thermal effects, which lead to significant degradation and variations among energy-storage units. After that—taking into account energy storage unit run-time effects—the proposed system-level ESS run-time charge capacity characterizes the ESS run-time charge efficiency. Due to the fact that large-scale ESS has complex connectivity interrelations, Wu et al. have developed a matrix differential equation for run-time charge capacity, as follows:

$$\frac{d\mathbf{C}_{e[N\times N]}(t)}{dt} = \mathbf{K}_{[N\times N]} \cdot \mathbf{V}_{[N\times N]} \cdot I \cdot U(t)$$

$$+ \mathbf{C}_{e[N\times N]}(t_0) \cdot \frac{d\mathbf{\Omega}_{[N\times N]}(t)}{dt}, \qquad (7.2)$$

where matrix $\mathbf{C}_{e[N\times N]}(t)$ is a diagonal matrix that models the run-time charge capacities of the N energy-storage units. Matrix $\mathbf{K}_{[N\times N]}$ models the ESS topology and the corresponding current distribution I among the N units. Diagonal matrix $\mathbf{V}_{[N\times N]}$ represents the ESS output voltage among the N units. $U(t)$ is a step function. Diagonal matrix $\mathbf{\Omega}_{[N\times N]}$ models the run-time aging of individual units, which is a function of $\mathbf{C}_{e[N\times N]}(t)$. Note that the variable $\mathbf{\Omega}_{[N\times N]}$ is a matrix form of the aging factor ω, as follows.

$$\mathbf{\Omega}(t) = \{\frac{1}{\Xi} \cdot (1 - e^{\frac{\Psi(t)}{\Lambda}})\}^{\frac{1}{\Theta}}, \text{ where} \qquad (7.3)$$

$$\mathbf{\Psi}(t) = \mathbf{K} \cdot \mathbf{R} \cdot I \cdot -(\mathbf{V}_{oc}(t) - \mathbf{V}_c),$$

Where $\mathbf{R}$, $\mathbf{V}_{oc}$, $\mathbf{V}_c$, Ξ, and Θ are the matrix form of resistance r, open-circuit voltage V_{oc}, cutoff voltage V_c, ξ, and κ, specifically. The proposed run-time charge capacity and aging model capture the characteristics of the ESS run-time aging effect on the capacity degradation phenomenon and run-time charge efficiency by considering the ESS connectivity information.

7.4   ESS-Aware Motorists Driving Analysis

(P)HEVs represent a useful technology to reduce the dependence of transportation on petroleum and reduce the prevalence of greenhouse emissions. The distinguishing feature of a (P)HEV is the ability of the vehicle to charge the energy storage system (ESS) from the electrical grid, which has a stronger link with the driving habits of motorists compared to a conventional vehicle. For instance, the heterogeneous driving habits of motorists significantly leads to differences in the operation of a (P)HEV that affect the corresponding ESS performance as well as having other environmental impacts. Therefore, accurately capturing and analyzing motorist-vehicle interaction at runtime has become a major bottleneck. To address this issue, this section discusses the impacts of performing driving analyses of ESS-aware motorists and presents an overview of existing studies in this area.

7.4.1 Impacts of ESS-Aware Driving Behavior Analysis

The run-time operation of the (P)HEV is tied to the driving behavior pattern of the vehicle's owner. Different motorists drive their vehicles differently. The ESS-aware driving behavior analysis is aimed at collection and analysis of data from the HEV under different operational modes to estimate the impact of driving behavior patterns on vehicle performance.

Due to the fact that (P)HEV performance substantially depends on the operation of the vehicle, existing researchers classify three different modes for use during a trip [58]:

1. **Charge Sustain (CS) mode:** In this mode, the (P)HEV acts as a conventional Hybrid Electric Vehicle, with charge and discharge cycles, and tries to sustain the SOC level. When a driver applies pedal pressure to cause acceleration, the battery provides necessary auxiliary power, and in this case, discharge current is observed. When brake is applied and the vehicle is decelerating or standing still, the batteries recharge and charge current is observed. Therefore, driving behavior primarily determines the current profile.
2. **Charge Deplete (CD) mode:** In this mode, once the vehicle is fully charged, it can be operated almost exclusively (except during hard acceleration) on electric power until its battery state of charge is depleted to a predetermined level, at which time the vehicles internal combustion engine (ICE) or fuel cell will be engaged.
3. **Blended mode:** Blended mode is a special kind of charge-depleting mode and usually employed by vehicles, Toyota Prius for instance, which do not have enough electric power to sustain high speeds without the help of the internal combustion portion of the power train. In (P)HEV, when the speed is less than a preset value which is considered to be the speed below which the engine cannot operate steadily, vehicle will only use electric power no matter how the driving behavior is, while the engine idles. This is called EV mode by Toyota, which is short for Electric Vehicle mode. This mode is more like standard CD mode except for the condition of speed limitation. At faster speed, the ICE will be used to provide power, while electric power can still be available in reserve.

Based on the different run-time operation modes, how a motorist drives has a significant impact on (P)HEV fuel efficiency, emissions, and ESS use. Two impacts of motorist driving behavior analysis are as follows:

- Different driving behavior directly affects the operation of the (P)HEV internal combustion engine, hence the impact on fuel efficiency and greenhouse gas emissions.
- Heterogeneous driving behavior of motorists affects the run-time charging–discharging of the ESS and its long-term lifetime.

7.4.2 Overview of ESS-Aware Driving Behavior Analysis

Considering the above impacts of driving behaviors of different motorists, heterogeneous driving behaviors have great impact on the vehicle economy and energy consumptions of different power split control strategies of (P)HEVs. Currently, various research works have focused on the different road conditions, traffic conditions, and vehicle energy consumption analyses of (P)HEVs.

- **Standard Driving Cycles Analysis:** Ericsson has demonstrated the factors of driving behavior and estimated their influence on the greenhouse emissions and fuel consumption [59]. Utilizing factorial analysis on driving patterns, Ericsson has extracted the independent factors that describe driving behaviors. Further, this study investigates which independent factors have significant effect on fuel consumption and greenhouse emissions. According to a certain speed profile, the study built performance models of the greenhouse emissions of HC, NO_x, and CO_2 and the engine system using an engine map and other vehicle parameters of the specified vehicle.

  Lin et al. have used an Artificial Neural Network (ANN) technique to recognize driving patterns [60]. Meanwhile, Lin et al. also designed a rule-based control strategy [61] to optimize parameters for control on each RDP (Representative Driving Pattern) and presented how heterogeneous driving behavior can result in different driving cycles under different compositions of modal events (i.e., cruise, idle, acceleration, and deceleration). Based on driving cycles, this study has analyzed the variance on average speed and average road power, statistically.

  Dembski et al. have studied statistical analysis and designed a clustering technique for standard driving cycles. This technique [62] has separated the driving cycles into segments based on statistical information. Meanwhile, this study has proposed a technique to reproduce the driving cycle by choosing the segments from the existing database.

  Ganji et al. have analyzed the standard driving cycles under different powertrain configurations (i.e., conventional, series hybrid, parallel hybrid, and series-parallel hybrid vehicles) to simulate the real-world conditions for vehicle performance analysis [63]. This study has shown that using two energy sources (i.e., fuel and ESS) in the propulsion system allows for a very diverse set of powertrain configurations.

– **Real-World Driving Cycles Analysis:**
Gong et al. have presented a systematic analysis using a clustering technique for real-world driving profiles and developed a driving pattern recognition algorithm based on the results of the clustering [64]. Further, the study built a Markov-chain model for the stochastic velocity reproduction for different driving behaviors. The final goal of this study is to estimate the impact of (P)HEVs on electric energy consumption and fuel consumption.

Li et al. have studied real-world driving analysis using a systematic approach, which leverages multi-modality driver-vehicle information to identify the corresponding operation modes under specific driving behaviors [65]. This study has modeled (P)HEV ESS usage, fuel consumption, and CO_2 emissions, enabling comprehensive and quantitative analysis of (P)HEV economic and environmental impacts under motorist-specific driving behavior.

7.5   ESS Architecture

Commonly, a large-scale ESS consists of thousands of energy storage units. Heterogeneous characteristics of energy storage units cause the whole ESS to be charged or discharged differently, even though all units are initially identical. Thus the capacity and lifetime of a large-scale ESS are determined by the weakest energy storage unit. Manufacturing tolerance, along with heterogeneous run-time usage and the ambient environment, lead to significant degradations and variations among individual energy storage units which results in serious ESS lifetime reliability concerns. Therefore, an active battery management system (BMS) is a must to monitor, control, and balance the pack of ESS. In the following sections, we will introduce the ESS architecture overview and the challenges of architecture design.

7.5.1   ESS Architecture Overview

Considering the high power/energy requirement of (P)HEVs, an ESS consists of a large number of energy storage units (more than 1,000) connected in parallel and series and controlled by a battery management system (BMS). The BMS is defined as an optimal control system to guarantee that the ESS supplies optimum energy to power the (P)HEV and guarantee that the risk of failure on the ESS is minimized. This is obtained by monitoring and controlling the ESS run-time charging–discharging current, temperature, as well as output voltage. Therefore, the fundamental undertaking of the BMS [66] is shown as follows:

- **Effective controlling and monitoring:** Due to the ESS capacity fading caused by heterogeneous manufacture tolerance and the heterogeneous run-time usage, the BMS must suitably control and monitor charging/discharging of the ESS, including tracking the state of charge (SOC) and the state of health (SOH) of ESS, with practically no overcharging or over discharging, so as to satisfy the run-time performance and the long-term cycle life of the ESS.
- **Practical powering:** Due to the differences in operational voltage and current between the large-scale ESS and the electric-drive propulsion components, the BMS powers the ESS to supply the minimum voltage and current to load, using DC/DC conversion, in order to achieve a longer run-time of the (P)HEV.

A general ESS architecture consists of a large number of energy storage units, a BMS, a DC/DC converter, and electric-drive propulsion components. The intelligence in the ESS architecture is included in two functions. One is the monitoring function, which involves the measurement of the characteristics of the ESS such as voltage, current, temperature, SOC and so on; the other is the controlling function, which acts on the charging–discharging of the ESS based on the measured variables. According to the literature [66, 67], the general ESS architecture has two types. One type is a central control scheme, which is responsible for monitoring and controlling all energy storage units. The central control scheme is straightforward to implement, but does not scale well; i.e., this architecture is not energy efficient when the number of energy storage units increase, resulting in large management latency. The other is the distributed control scheme, which is an individual control module that takes responsibility for each energy storage unit independently. The distributed architecture monitors and controls the energy storage unit efficiently. However, as the numbers of energy storage unit increase, the cost of the distributed BMS grows rapidly. Therefore, we need the smart ESS architecture to maximize the system performance and reliability while with minimum cost.

7.5.2   ESS Architecture Challenges

The ESS architecture design needs to jointly consider the cost of monitoring and controlling the components, the ESS run-time performance, and the long-term cycle life. More specifically,

- The performance (e.g., energy capacity) and cycle life of an ESS are constrained by the weakest battery cell. Due to the limited capabilities of monitoring cells and system SOC and SOH over high-voltage boundaries, existing BMS offers limited control over individual cells. As a result, mismatches among cells—due to manufacturing tolerance, temperature gradients across the pack, and mismatched degradation over cycle and calendar life—can lead to overcharging or excessive discharge of individual unit, resulting in overall system performance degradation and severe cycle life limits. This problem becomes increasingly worse with the increase of the number of energy storage units.

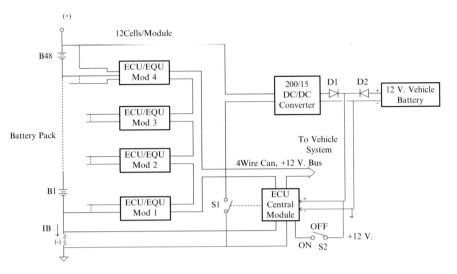

Fig. 7.2 A prototypical modular system for a Li-ion pack [67]

- Even with the help of the BMS, the string length (i.e., the number of cells or modules in series) is limited. An additional bidirectional high voltage DC-DC converter is therefore required between the battery pack and the electric-drive propulsion components, which regulates the propulsion DC bus voltage V_{DC} but also increases the overall ESS complexity and cost.
- Existing BMS implementations support same-technology and same-spec battery cells only. They do not allow heterogeneous storage technology integration (e.g., power cells, energy cells, and ultracapacitors) or storage units with different design specs, therefore seriously limiting the overall ESS design optimization space.

7.5.3 Existing ESS Architectures

Currently, many studies have focused on the ESS architecture design to address the above issues. Stuart et al. proposed a modular-based ESS architecture [67], which consists of four local modules and one central module to monitor battery cells, as shown in Fig. 7.2. This modular-based ESS architecture has reduced the wiring latency and has better energy-efficiency than the centralized ESS architecture. Also, the modular-based ESS architecture provides three advantages. Firstly, it has developed an accurate circuit to measure the voltage of each module. Secondly, it has designed the charge measurement circuit to assess the run-time charge–discharge current waveform. Thirdly, it has proposed a relative current routing

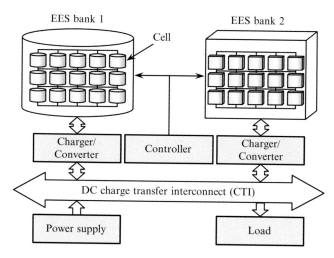

EES bank 1   EES bank 2

Cell

Fig. 7.3 Hybrid electrical energy storage architecture [68]

circuit to boost the charge on low voltage. However, since the modular-based ESS architecture requires a local module to monitor and control cells—along with a global module to synthesize local modules—the cost of the whole ESS architecture increases as the cells increases.

To address these issues, Kim et al. proposed a hybrid electrical energy storage (HEES) system architecture [68] to maximize the overall cycle efficiency. A HEES system is an EES that consists of two or more heterogeneous EES elements to leverage the advantages of each type. The HESS architecture consists of heterogeneous EES banks, and each bank is composed of homogeneous EES elements cells, which is shown in Fig. 7.3.

Energy is transferred between banks, from a bank to the load, or from the power supply to a bank over a DC charge transfer interconnect (CTI). Power converters are placed in between the CTI and EES banks for regulating voltage and/or current. A bank is typically organized as a two-dimensional array structure with a number of parallel and series connections of cells in order to provide more output power, larger energy capacity, or higher voltage level. The power capacity and voltage rating are determined by the number of parallel and series connections.

This study introduced a dynamic HEES bank reconfiguration method considering power conversion issues as the first step in order to realize higher cycle efficiency and storage capacity utilization with minimum HEES system cost. Kim et al. have applied the proposed the reconfiguration technique to ultracapacitor banks, and demonstrated that the cycle efficiency could be improved by up to 108% for a constant given power profile, whereas the pulse duty cycle was improved by up to 127% for a high-current pulsed power profile.

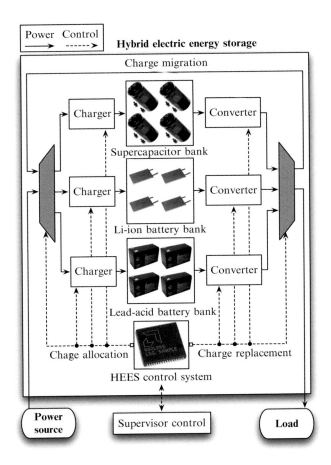

Fig. 7.4 Architecture of the hybrid EES system [32]

In addition to the existing works, Pedram et al. proposed a memory hierarchy structure to build a heterogeneous hybrid ESS system architecture [32] to organize and allocate various types of energy storage units, which is shown in Fig. 7.4. The heterogeneous hybrid ESS architecture consists of heterogeneous ESS elements, with each unit connected through DC-DC converters. In this architecture, it has adopted a flat topology, which means a single controller is responsible for monitoring and controlling all energy storage units. In comparison to the existing flat structure, the memory-based flat structure combines the flat physical topology and the hierarchical logical structure to take the advantage of shortening charge transfer paths, reducing the losses in charge allocation processes, and allowing any combination of charge allocation, migration and replacement policies.

7.6 System Optimization and Control

Several researchers have developed numerous optimization and control strategies for ESSs. This section presents the overview of such optimization and control strategies. These strategies can be classified into two application areas: one is the portable embedded system application, another is the (P)HEV application.

7.6.1 Optimization and Control Strategies in Portable Embedded Systems

Considering the portable embedded system with small size, such as mobile devices, researchers have focused on maximization of ESS short-term lifetime with single battery cell and designed plentiful strategies of optimization and control for the ESS.

Because of the high power density of the ultracapacitor, recent studies have started to consider hybrid energy storage technology, named battery-ultracapacitor integration, which shows a promising way to improve the performance of the ESS [31, 32]. During 2006–2012, Pedram et al. published several papers that discussed strategies of optimization and control for the ESS [8, 69–72].

Peng et al. have described the problem of maximizing the capacity utilization of the battery power source in the portable embedded system under latency and loss rate constraints. Based on this problem, this study has built a battery-aware power management technique using the continuous- time Markovian Decision Processes (MDP) and stochastic networks [8] to capture the current rate-capacity feature and capacity recovery feature of the battery cell. This study aims to maximize the battery short-term service lifetime while satisfying the given service timing constraints. In comparison to the existing heuristic techniques for battery management, the proposed algorithm achieves 17% improvement in average energy delivered per energy storage unit weight.

Shin et al. have proposed a battery-ultracapacitor hybrid architecture [69] to maximize the deliverable energy density under different constraints for portable electronics. This study also has developed a design space exploration algorithm to maximize the deliverable energy density for a given charging–discharging current profile. The proposed hybrid optimization architecture can reduce the capacity loss due to the highly fluctuating charging–discharging current profile. The proposed optimization algorithm achieves 7.7% improvement in deliverable energy density over the conventional parallel connection of battery and ultracapacitor.

Wang et al. have developed the optimization technique of charge migration for hybrid electric energy storage systems that finds the best migration efficiency [70]. This study has defined time-unconstrained and time-constrained charge migration problems and provided a systematic derivation of the optimal charge migration. Compared with the baseline charge migration technique, the proposed optimization method enhances the global migration efficiency up to 51.3%.

Xie et al. have presented the charge allocation problem and developed the systematic technique for maximum charge allocation efficiency [71]. Based on the generalized hybrid electric energy storage (HEES) architecture, this study has built the corresponding electrical circuit models to describe the chargers and banks. Further, utilizing the mixed integer nonlinear optimization technique, this study designed an optimal algorithm to find the global charge allocation efficiency for long-term charge allocation processes. After that, Xie et al. have described the global charge replacement (GCR) optimization problem for HEES. Meanwhile, this study has designed an algorithm to achieve the near-optimal GCR control policy [72] under many constraints (i.e., energy reservation, converter efficiency, rate capacity effect, and self-discharge rates). In comparison to the baseline setup, the proposed near-optimal GCR control algorithm achieves 42.8% improvement.

In addition, Mirhoseini et al. have developed a hybrid battery-ultracapacitor power supply optimization [73] to improve the lifetime of portable systems. Further, Mirhoseini et al. have proposed an optimal management strategy [74] for the hybrid battery-ultracapacitor system.

7.6.2 Optimization and Control Strategies in the (P)HEV

Although the aforementioned methods for portable embedded systems offer optimal solutions for single battery cells, they are challenging, if not impossible, to extend to the (P)HEV application. The (P)HEV consists of a large number of energy storage units (even more than 1,000) connected in parallel and serial to provide output voltage, driving current, and energy storage capacity to vehicles. Direct extensions of existing methods used for single battery cells to (P)HEVs are therefore computational prohibitive and also fail to characterize the interaction effect among the energy storage units (e.g., heterogeneous capacity). Moreover, existing methods, developed for the maximization of short-term lifetime, have ignored the long-term aging effect, which is, however, important for the energy storage system (ESS) cost of the (P)HEV.

From the perspective of the (P)HEV application, the energy storage technology and hybrid ESS design have drawn significant attention in recent years. Markel et al. have investigated (P)HEV technologies and pointed out the power and the energy capacity as the two critical factors in (P)HEV ESS design [1]. Rousseau et al. have presented an ESS design considering vehicle design, control strategies, and drive cycle [75], based on existing HEV configurations. Dumitrescu et al. have proposed a hybrid integration of NiMH battery and double-layer capacitor technologies [76]. This study has designed a technique for a power management system of loads with large peak-to-average power ratios. Smith has integrated fuel cells with ultracapacitors for ESS power and energy optimization [77]. Burke has analyzed the feasibility of incorporating ultracapacitors into electric vehicle battery systems [78]. Cooper et al. have developed a hybrid lead-acid battery-ultracapacitor ESS called UltraBattery, which demonstrated that both run-time power demand and lifetime can be enhanced [79]. In this study, the ultracapacitor acts as a buffer to share the discharging and charging currents with the lead-acid battery, thus enabling it to provide and absorb charge rapidly during vehicle acceleration and braking. Lukic et al. have summarized the different charge control strategies with different topologies of hybrid ESSs [80]. Garcia et al. have proposed the control strategy of charge allocation for the battery-ultracapacitor to coordinate the (P)HEV power demand [81]. The proposed control strategy in [81] regulates the output voltage based on the combination of two different energy storage technologies (i.e., battery-ultracapacitor), in order to keep stability. Zhou et al. have developed a run-time power management strategy [82] for multi-source (e.g., lead-acid battery and ultracapacitor) in the hybrid electric vehicle. In addition, Wu et al. have proposed a design framework that unifies design-time optimization and run-time control [83]. This study has distinguished itself from existing studies, not only because it minimizes the ESS cost—instead of the run-time energy consumption— by optimizing the ESS configurations, but also because it models several effects that are important to the ESS cost and lifetime. Targeting the ESS architecture, the design-time optimization is to quantitatively determine the ESS configuration, including the number of storage units, as well as the type and size of each unit, to minimize the system cost while ensuring the target lifetime for most (P)HEV vehicle drivers. In the (P)HEV application, the ESS system consists of $N = m \times n$ energy storage units organized in an $m \times n$ regular array. This means m modules connected in series and n units connected in parallel for each module. For each energy storage unit i, j with type $q_{i,j} \in Q$ and size $s_{i,j}$, the ESS cost is minimized while satisfying the usage demand for the target lifetime. Therefore, the ESS cost function is shown as follows:

$$cost(\mathbf{s}) \overset{\triangle}{=} \sum_{i=1}^{m} \sum_{j=1}^{n} c_{i,j}(s_{i,j}), \tag{7.4}$$

where the $c_{i,j}(s_{i,j})$ is the cost function for energy storage unit i, j with size $s_{i,j}$ and type $q_{i,j}$.

In order to meet the power demand, the energy demand, and the particular remaining capacity requirements, the constraints of this study is shown as follows:

$$\Pr[P^{ESS}(\mathbf{s}) \geq P^d \wedge E^{ESS}(\mathbf{s}) \geq E^d \wedge A^{ESS}(\mathbf{s}) \leq \eta] \geq \delta, \mathbf{s} \in \mathcal{S}. \tag{7.5}$$

The P^{ESS} and E^{ESS} are denoted as the power and energy capacities of the ESS at the end of the lifespan, separately. A^{ESS} represents the capacity aging effect of the ESS. P^d and E^d are separately denoted as the maximal power usage demand and energy demand among the motorists. η is an upper-bound of efficiency used for each energy storage unit. The $\delta \in [0, 1]$ is the statistical lifetime guarantee. A lifetime guarantee equal to 1 is equivalent to the deterministic optimization that targets at the worst-case scenarios of both the ESS manufacture process variations and the driving patterns (i.e., ESS run-time use) to meet a particular remaining capacity requirement. Overall, compared against the worst-case based Li-ion only ESS, the produced hybrid ESS designs reduce the system cost on average by 51.2% with only 5% lifetime guarantee loss.

7.7 Conclusions

(P)HEVs present an excellent opportunity to not only reduce transportation petroleum dependencies but also to curb greenhouse gas emissions. This chapter overviewed the challenges found in energy storage systems (ESS) design for the (P)HEV application area. Firstly, the performance metrics that drive ESS design were outlined (i.e., cost, energy density, peak power, run-time charge-cycle efficiency, and lifetime), along with the electrochemical effects of various energy storage technologies (e.g. Li-ion batteries and ultracapacitors). Next, ESS modeling methodologies—from the single ESS unit to large-scale ESS—were compared. Impacts of motorist driving behaviors and driving cycles were shown on ESS performance and the environment. ESS architectures, including hybrid electrical energy storage (HEES) architectures, were presented. Finally, (P)HEV optimization and control strategies were given for combining and overall balancing of all ESS performance metrics.

References

1. Markel T, Simpson A (2006) Cost-benefit analysis of plug-in hybrid electric vehicle technology. In: 22nd international electric vehicle symposium, Yokohama
2. Environmental assessment of plug-in hybrid electric vehicles, vol 1: nationwide greenhouse gas emissions. Electric Power Research Institute (EPRI), Palo Alto, CA, Tech. Rep. 1015325, July 2007
3. Samaras C, Meisterling K (2008) Life cycle assessment of greenhouse gas emissions from plug-in hybrid vehicles: implications for policy. Environ Sci Technol 42(9):3170–3176
4. Karplus V et al (2012) Should a vehicle fuel economy standard be combined with an economy-wide greenhouse gas emissions constraint? Implications for energy and climate policy in the united states. Energy Econ 36: 322–333
5. Miller JM (2009) Energy storage system technology challenges facing strong hybrid, plug-in and battery electric vehicles. In: IEEE vehicle power propulsion dsonference, Dearborn, pp 4–10
6. Pang C, Dutta P, Kezunovic M (2012) Bevs/phevs as dispersed energy storage for v2b uses in the smart grid. IEEE Trans Smart Grid 3(1):473–482
7. Lahiri K, Raghunathan A, Dey S (2004) Efficient power profiling for battery-driven embedded system design. IEEE Trans Comput-Aided Des Integr Circuits Syst 23(6):919–932
8. Rong P, Pedram M (2006) Battery-aware power management based on markovian decision processes. IEEE Trans Comput-Aided Des Integr Circuits Syst 25(7):1337–1349
9. Li Y et al (2012) An energy efficient solution: Integrating plug-in hybrid electric vehicle in smart grid with renewable energy. In: IEEE conference on computer communications workshops (INFOCOM WKSHPS), Orlando, 2012. IEEE, pp 73–78
10. Baisden A, Emadi A (2004) Advisor-based model of a battery and an ultra-capacitor energy source for hybrid electric vehicles. IEEE Trans Veh Technol 53(1):199–205
11. Lukic SM et al (2008) Energy storage systems for automotive applications. IEEE Trans Ind Electron 55(6):2258–2267
12. Cao J, Emadi A (2012) A new battery/ultracapacitor hybrid energy storage system for electric, hybrid, and plug-in hybrid electric vehicles. IEEE Trans Power Electron 27(1):122–132

13. Maksimovic D, Zane R, Erickson R (2009) Multi-cell battery systems. University of Colorado at Boulder (UCB), Invention Disclosure
14. US Advanced Battery Consortium. http://www.uscar.org/
15. Pesaran A, National Renewable Energy Laboratory (U.S.) et al (2009) Battery requirements for plug-in hybrid electric vehicles–analysis and rationale. National Renewable Energy Laboratory, Golden
16. Vazquez S et al (2010) Energy storage systems for transport and grid applications. IEEE Trans Ind Electron 57(12):3881–3895
17. Martin TL (1999). Balancing batteries, power, and performance: system issues in Cpu speed-setting for mobile computing. Ph.D. Dissertation, Carnegie Mellon University, Pittsburgh, PA, USA
18. Panigrahi D et al (2001) Battery life estimation of mobile embedded systems. In: Proceedings of the 14th IEEE/ACM international conference on VLSI design, San Diego
19. Li K et al (2010) Large-scale battery system modeling and analysis for emerging electric-drive vehicles. In: ACM proceedings of the 2010 international symposium on low power electronics and design (ISLPED), Austin
20. Vetter J et al (2005) Ageing mechanisms in lithium-ion batteries. J Power Sources 147(1):269–281
21. Toyota PHEV technologies. http://www.toyota.com
22. Seo H et al (2010) Power quality control strategy for grid-connected renewable energy sources using pv array and supercapacitor. In: International conference on electrical machines and systems (ICEMS) 2010. IEEE, Incheon, Korea (South), pp 437–441
23. Rong P, Pedram M (2006) An analytical model for predicting the remaining battery capacity of lithium-ion batteries. IEEE Trans Very Larg Scale Integr Syst 14(5):441–451
24. Hung S, Hopkins D, Mosling C (1993) Extension of battery life via charge equalization control. IEEE Trans Ind Electron 40(1):96–104
25. Moawad A et al (2009) Impact of real world drive cycles on phev fuel efficiency and cost for different powertrain and battery characteristics. In: International battery, hybrid and fuel cell electric vehicle symposium, Stavanger
26. Shiau CSN et al (2009) Impact of battery weight and charging patterns on the economic and environmental benefits of plug-in hybrid vehicles. Energy Policy 37(7):2653–2663
27. Fuller T, Doyle M, Newman J (1994) Simulation and optimization of the dual lithium ion insertion cell. J Electrochem Soc 141(1):1–10
28. Johnson VH (2002) Battery performance models in ADVISOR. J Power Sources 110: 321–329
29. Rao V et al (2005) Battery model for embedded systems. In: 18th international conference on VLSI design, Kolkata, 2005. IEEE, pp 105–110
30. Rakhmatov D, Vrudhula S, Wallach D (2002) Battery lifetime prediction for energy-aware computing. In: ISLPED '02: proceedings of the 2002 international symposium on Low power electronics and design, New York, pp 154–159
31. Lukic SM et al (2006) Power management of an ultra-capacitor/battery hybrid energy storage system in an HEV. In: IEEE vehicle power propulsion conference, Windsor, United Kingdom, pp 1–6
32. Pedram M et al (2010) Hybrid electrical energy storage systems. In: ACM/IEEE international symposium on low-power electronics and design (ISLPED), Austin, 2010. IEEE, pp 363–368
33. Du Pasquier A et al (2003) A comparative study of li-ion battery, supercapacitor and nonaqueous asymmetric hybrid devices for automotive applications. J Power Sources 115(1):171–178
34. Lahiri K et al (2002) Battery-driven system design: a new frontier in low power design. In: Design automation conference, 2002. Proceedings of ASP-DAC 2002. 7th Asia and South Pacific and the 15th international conference on VLSI design, Bangalore
35. Khateeb SA et al (2006) Mechanical-electrochemical modeling of Li-ion battery designed for an electric scooter. J Power Sources 158(1):673–678
36. Ricketts B, Ton-That C (2000) Self-discharge of carbon-based supercapacitors with organic electrolytes. J Power Sources 89(1):64–69

37. Conway B (1999) Electrochemical supercapacitors: scientific fundamentals and technological applications. Springer, New York
38. Diab Y et al (2009) Self-discharge characterization and modeling of electrochemical capacitor used for power electronics applications. IEEE Trans Power Electron 24(2):510–517
39. Santhanagopalan S et al (2005) Review of models for predicting the cycling performance of lithium ion batteries. J Power Sources 156: 620–628
40. Kazuo O et al (2003) Study on heat generation behavior of small lithium-ion secondary battery. J Electrochem Soc 150(3):A285–A291
41. Newman JS (1999) FORTRAN programs for simulation of electrochemical systems. Available: http://www.cchem.berkeley.edu/~jsngrp/.
42. Smith K, Rahn C, Wang C (2010) Model-based electrochemical estimation and constraint management for pulse operation of lithium ion batteries. IEEE Trans Control Syst Technol 18(3):654–663
43. Doyle M, Fuller T, Newman J (1993) Modeling of galvanostatic charge and discharge of the lithium/polymer/insertion cell. J Electrochem Soc 140(6):1526–1533
44. Rakhmatov D, Vrudhula S, Wallach D (2003) Model for battery lifetime analysis for organizing applications on a pocket computer. IEEE Trans Very Larg Scale Integr Syst 11(6):1019–1030
45. Gao L, Liu S, Dougal R (2002) Dynamic lithium-ion battery model for system simulation. IEEE Trans Compon Packag Technol 25(3):495–505
46. Hageman S (1993) Simple pspice models let you simulate common battery types. Edn-Boston Denver 38:117–117
47. Yann Liaw B et al (2004) Modeling of lithium ion cellsa simple equivalent-circuit model approach. Solid State Ion 175(1):835–839
48. Lee S et al (2008) State-of-charge and capacity estimation of lithium-ion battery using a new open-circuit voltage versus state-of-charge. J Power Sources 185(2):1367–1373
49. Chiasserinia C, Rao R (2000) Stochastic battery discharge in portable communication devices. IEEE Aerosp Electron Syst Mag 15(8):41–45
50. Chiasserini C, Rao R (1999) A model for battery pulsed discharge with recovery effect. In: Wireless communications and networking conference (WCNC), 1999. IEEE, New Orleans, LA, pp 636–639
51. Chiasserini C, Rao R (2001) Improving battery performance by using traffic shaping techniques. IEEE J Sel Areas Commun 19(7):1385–1394
52. Chiasserini C, Rao R (2001) Energy efficient battery management. IEEE J Sel Areas Commun 19(7):1235–1245
53. Manwell J, McGowan J (1994) Extension of the kinetic battery model for wind/hybrid power systems. In: Proceedings of EWEC, Thessaloniki, Greece, pp 284–289
54. Nelson, Amine K (2007) Advanced lithium-ion batteries for plug-in hybrid-electric vehicles. In: 23rd international electric vehicle symposium (EVS23), Argonne National Laboratory, Lemont
55. A. N. Laboratory, PSAT (Powertrain Systems Analysis Toolkit). http://www.transportation.anl.gov/
56. Kroeze R, Krein P (2008) Electrical battery model for use in dynamic electric vehicle simulations. In: Power electronics specialists conference (PESC), 2008. IEEE, Rhodes, Greece, pp 1336–1342
57. Wu J et al (2011) Large-scale battery system development and user-specific driving behavior analysis for emerging electric-drive vehicles. Energies 4:758–779
58. Midlam-Mohler S et al (2009) Phev fleet data collection and analysis. In: Vehicle power and propulsion conference (VPPC'09), 2009. IEEE, Dearborn, MI, pp 1205–1210
59. Ericsson E (2001) Independent driving pattern factors and their influence on fuel-use and exhaust emission factors. Transp Res Part D: Transp Env 6(5):325–345
60. Lin C et al (2002) Control of a hybrid electric truck based on driving pattern recognition. In: Proceedings of the 6th international symposium on advanced vehicle control, Hiroshima
61. Lin C et al (2004) Driving pattern recognition for control of hybrid electric trucks. Veh Syst Dyn 42(1–2):41–58

62. Dembski N et al (2005) Development of refuse vehicle driving and duty cycles. SAE Trans 114(2):90–102
63. Ganji B, Kouzani A, Trinh H (2010) Drive cycle analysis of the performance of hybrid electric vehicles. In: Life system modeling and intelligent computing. Springer, New York, pp 434–444
64. Gong Q et al (2010) Statistical analysis of phev fleet data. In: Vehicle power and propulsion conference (VPPC), 2010. IEEE, pp 1–6
65. Li K et al (2012) Personalized driving behavior monitoring and analysis for emerging hybrid vehicles. In: Pervasive computing. Springer, New York, pp 1–19
66. Bergveld H, Kruijt W, Notten P (2002) Battery management systems: design by modelling, vol 1. Springer, Boston
67. Stuart T et al (2002) A modular battery management system for hevs. In: Proceedings of the SAE future car congress (Paper number 2002-01-1918), Arlington
68. Kim Y et al (2010) Balanced reconfiguration of storage banks in a hybrid electrical energy storage system. In: Proceedings of the international conference on computer-aided design. IEEE, San Jose, CA, pp 624–631
69. Shin D et al (2011) Constant-current regulator-based battery-supercapacitor hybrid architecture for high-rate pulsed load applications. J Power Sources 205:516–524
70. Wang Y et al (2011) Charge migration efficiency optimization in hybrid electrical energy storage (hees) systems. In: ISLPED'11, Fukuoka, pp 103–108
71. Xie Q et al (2011) Charge allocation for hybrid electrical energy storage systems. In: Proceedings of the 9th international conference on hardware/software codesign and system synthesis (CODES+ISSS), Taipei, 2011. IEEE, pp 277–284
72. Xie Q et al (2012) Charge replacement in hybrid electrical energy storage systems. In: 17th Asia and South Pacific design automation conference (ASP-DAC), Sydney, 2012. IEEE, pp 627–632
73. Mirhoseini A, Koushanfar F (2011) Hypoenergy hybrid supercapacitor-battery power-supply optimization for energy efficiency. In: Design, automation & test in Europe conference & exhibition (DATE), Grenoble, 2011. IEEE, pp 1–4
74. Mirhoseini A, Koushanfar F (2011) Learning to manage combined energy supply systems. In: International symposium on low power electronics and design (ISLPED) 2011. IEEE, Fukuoka, pp 229–234
75. Rousseau A et al (2007) Research on phev battery requirements and evaluation of early prototypes. In: 7th advanced automotive battery conference. Long Beach, CA
76. Roman Dumitrescu CR, Gausemeier J (2009) Design methodology of a combined battery-ultracapacitor energy storage unit for vehicle power management. In: 10th international workshop on research and education in mechatronics. Glasgow, UK
77. Smith R (2004) Fuel cells and ultracapacitors. In: Advanced capacitor world summit 2004. Washington, DC
78. Burke AF (2007) Batteries and ultracapacitors for electric, hybrid, and fuel cell vehicles. Proc IEEE 95(4):806–820
79. Cooper et al A (2009) The ultrabattery–a new battery design for a new beginning in hybrid electric vehicle energy storage. J Power Sources 188(2):642–649
80. Lukic et al S (2008) Energy storage systems for automotive applications. IEEE Trans Ind Electron 55(6):2258–2267
81. Garcia F, Ferreira A, Pomilio J (2009) Control strategy for battery-ultracapacitor hybrid energy storage system. In: Twenty-fourth annual IEEE applied power electronics conference and exposition (APEC), 2009. IEEE, Washington, DC, pp 826–832
82. Zhou Z et al (2011) Power management of passive multi-source hybrid electric vehicle. In: Vehicle power and propulsion conference (VPPC), 2011. IEEE, Chicago, IL, pp. 1–4
83. Wu J et al (2012) Large-scale energy storage system design and optimization for emerging electric-drive vehicles. IEEE Trans Comput-Aided Des Integr Circuit Syst 32:325–338

Chapter 8
Sensor Network Protocols for Greener Smart Environments

Giacomo Ghidini, Sajal K. Das, and Dirk Pesch

8.1 Introduction

A wireless sensor network (WSN), depicted in Fig. 8.1, consists of a set of sensor nodes, small battery-powered computing devices connected via a multi-hop wireless network [2, 40]. These nodes, also called sensors or motes, measure physical quantities of the surrounding environments using on-board sensors. Thanks to advances in micro-electronic mechanical systems (MEMS), there exist small form-factor sensors to measure a wide array of physical quantities: from temperature to humidity, from strain to electromagnetic field, to name a few. These analog data are then converted to digital values and relayed to the base station, or sink, along a multi-hop route formed by the motes in the wireless network. These data may be stored in memory at the source node, en route, or the base station, and processed by these nodes, in order to remove redundancy or noise, add error correction, or verify authenticity. Once the data reach the base station, they may be used to monitor the environment in which the WSN is deployed, build a model thereof, and make decisions as to what actions need be taken by actuators or humans.

Given their nature as a direct connection between the physical and cyber worlds, WSNs have potential applications in many diverse fields, and are one of the fundamental components to develop greener computing systems. In fact, one can imagine that most problems humankind is facing could be addressed in a more precise and (energy) efficient manner, if accurate data from the physical environment (be it a forest hosting an endangered species, a crop needing the

G. Ghidini (✉) • S.K. Das
Center for Research in Wireless Mobility and Networking, The University of Texas
at Arlington, Arlington, TX 76019, USA
e-mail: giacomo@uta.edu; das@uta.edu

D. Pesch
Cork Institute of Technology, Cork, Ireland
e-mail: dirk.pesch@cit.ie

P.P. Pande et al. (eds.), *Design Technologies for Green and Sustainable Computing Systems*,   205
DOI 10.1007/978-1-4614-4975-1_8, © Springer Science+Business Media New York 2013

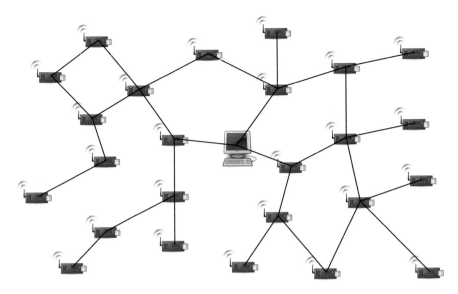

Fig. 8.1 A wireless sensor network consisting of motes connected to a base station via a multi-hop wireless network

appropriate quantity of water and fertilizers, or a freeway network often jammed by traffic) were available. So far, WSNs have been applied in several areas, including natural environments [32], rural areas [31], and urban domains [21]. Constraints on the motes due to the current technology, such as battery lifetime, size of the devices, and manufacturing costs, may temporarily delay the widespread application of WSNs to certain domains and scenarios. While WSNs are a necessary component towards more energy efficient systems in many domains, it is also important that the WSN protocols and applications be designed with energy efficiency as one of the driving objectives in order to make the overall system greener.

In this chapter, we aim to provide an overview of state-of-the-art algorithms and protocols for wireless sensor networks with a special focus on the communication stack. In all distributed systems, the protocols for the communication stack play a vital role by enabling interoperability of different subsystems. In wireless sensor networks, the communication stack is even more important because it accounts for the largest share of energy consumption in many application scenarios, and thus well-designed, energy-efficient protocols have a strong impact on the overall lifetime of the WSN.

In the past decade, several organizations, including the Institute of Electrical and Electronics Engineers (IEEE) and the Internet Engineering Task Force (IETF), have defined standards for physical layer, MAC layer, and network layer. In particular, we observe a trend towards the adoption of IEEE 802.15.4 [19] for physical (PHY) and medium access control (MAC) layers, and IETF 6LoWPAN [18,27] as the IPv6 protocol at the network layer. Standards for routing and ReSTful communication are also being proposed within the IETF. The IETF Routing over Low Power and Lossy

Fig. 8.2 Comparison of the
communication stacks used in
the Internet and in WSNs

Layer	Internet	WSNs
Application	HTTP	IETF CoAP
Transport	TCP	UDP
Network	IPv6	IETF 6LoWPAN / IETF RPL
Link (MAC)	IEEE 802.11	IEEE 802.15.4

Networks (RoLL) working group (WG) is developing the Routing Protocol for
LLNs (RPL) [38], a standard for routing in WSNs, while the Constrained ReSTful
Environments (CoRE) [20] is developing the Constrained Application Protocol
(CoAP), a standard for ReSTful communication with WSNs. Figure 8.2 displays
these protocols side-by-side with the corresponding ones already being used in the
Internet.

We argue that it is important to provide an overview of the existing and proposed
standards and their implementations not limiting the analysis to one layer, but rather
discussing them as part of this developing communication stack for WSNs. To this
extent, in our presentation of state-of-the-art solutions and proposed standards we
attempt to bring to the foreground the interdependencies between different layers
and the implications that design decisions at one layer have on the performance
at other ones. Ultimately, our analysis of the communication stack is aimed to
help researchers who are new to the area of WSN communications understand its
overall functioning, while also offering to more seasoned researchers an insight into
protocols at different layers and the interplay among them.

The rest of the chapter is organized as follows. In Sect. 8.2, we review the
different classes of MAC protocols, and describe the major features of IEEE
802.15.4. In Sect. 8.3, we present 6LoWPAN and RPL, respectively the standards
for IPv6 and routing in WSNs, and survey implementations and results. In Sect. 8.4,
we summarize the major features of CoAP, and then analyze recent evaluations of
the protocol. Finally, we draw our conclusions in Sect. 8.6.

8.2 MAC Layer

In a WSN, the MAC layer plays a vital role as it enables the actual communication
among nodes over a common medium. As such, a MAC protocol is often evaluated
along several dimensions, including delay, throughput, and energy efficiency.
However, constraints such as scarce battery capacity, limited computational power,
and small memory size, make the design of MAC protocols that can perform well
with respect to the performance metrics very difficult. Finally, the diversity of traffic
patterns generated by applications in diverse domains further complicate the design
of a general, effective, and efficient MAC protocol for WSNs.

Likely because of the challenges discussed above, researchers have dedicated
a lot of efforts to developing MAC protocols that meet all the requirements.

As a results, dozens, if not hundreds, of MAC protocols have been proposed by the research community in the past 15 years. In the past, and to a certain extent, still today, authors tried to classify MAC protocols based on the technique that they employ to coordinate access to a common medium by multiple nodes. These classifications are often variations of the one considering reservation-based protocols, contention-based protocols, and hybrid solutions. In this classification, reservation-based protocols usually feature some form of time-division multiple access (TDMA) and/or frequency-division multiple access (FDMA), while contention-based protocols are built around ALOHA or carrier-sense multiple access (CSMA), and hybrid protocols feature a mix of the two. The requirement for knowledge of the topology and strict synchronization are among the major drawbacks of the first class of protocols (i.e., reservation-based). Instead, the second class of protocols (i.e., contention-based) does not require this information, but experiences degraded performance in case of heavy traffic load and high energy consumption per bit (i.e., poor energy-efficiency) even in presence of light traffic loads.

8.2.1 MAC Protocol Classes

Recently, [4] proposes a classification of MAC protocols based on traffic patterns. The important assumption behind this classification is that the ultimate MAC protocol for WSNs with optimal performance for all traffic loads does not exist. Instead, there exist MAC protocols that are optimal for certain classes of WSN traffic.

The authors first summarize the causes of wasteful energy consumption at the MAC layer. In particular, they list: collisions, overhearing, overhead, and idle listening. Then they define three classes of traffic load: heavy, medium, and low. They observe how certain kinds of wasteful energy consumption are more likely to arise in presence of specific traffic loads. For instance, collisions are more common in case of heavy traffic, while idle listening is usually a cause of wasteful energy consumption in presence of light traffic. The authors then introduce three basic classes of MAC protocols: scheduled protocols, common active period-based protocols, and preamble sampling-based protocols. Hybrid protocols, such as IEEE 802.15.4, present features of different classes, and thus are grouped in a separate class.

8.2.1.1 Schedule-Based Protocols

In a scheduled protocol like TSMP [28], medium access is controlled by a schedule in the time and/or frequency domains. In a first version, communication links for each pair of neighboring nodes can be scheduled. Figure 8.3 portrays the communication between all pairs of nodes in a clique of four sensors. Alternatively, simply the senders or the receivers can be scheduled. The first option performs very

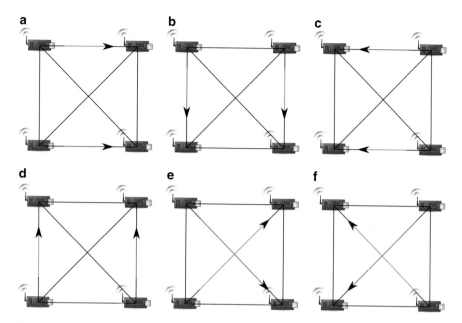

Fig. 8.3 Sample scheduled MAC protocol. *Black lines* depict communication links; *orange* and *blue arrows* depict scheduled communication on two different channels. (**a**) t = 0. (**b**) t = 1. (**c**) t = 2. (**d**) t = 3. (**e**) t = 4. (**f**) t = 5

well in presence of heavy traffic loads, but brings about major overhead since all pairs of neighboring nodes must be scheduled a slot for communication. Overhead is reduced by scheduling only senders or receivers, but other sources of wasteful energy consumption become relevant. In the solution where senders are scheduled, all neighbors have to listen to each sender, because the message may be addressed to them, thus resulting in overhearing. If receivers are scheduled instead, collisions may occur, so that this variant is not as effective in case of heavy traffic loads.

8.2.1.2 Common Active Period-Based Protocols

The next class of MAC protocols is that of those based on common active periods and targeted to medium traffic loads such as SMAC [39]. As depicted in Fig. 8.4, protocols in this class attempt to achieve a coarse synchronization between the active periods of neighboring nodes, so that they can communicate during these times, and turn off the radios at all other times. This is based on the assumption that a medium traffic load can be taken care of during a fraction of the node lifetime, and thus precious battery power can be saved by operating the radio only during these times. In the common active periods, nodes usually operate according to a contention-based mechanism such as CSMA to transmit their frames. In a common active period-based protocol, schedules are distributed so that all neighbors turn on

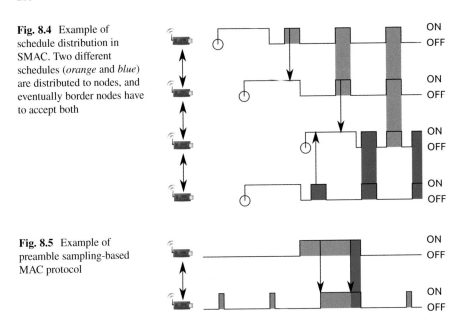

Fig. 8.4 Example of schedule distribution in SMAC. Two different schedules (*orange* and *blue*) are distributed to nodes, and eventually border nodes have to accept both

Fig. 8.5 Example of preamble sampling-based MAC protocol

their radios at the same time. As a result of this schedule distribution, clusters of nodes with the same schedule can be formed. As depicted in Fig. 8.4, in order to support inter-cluster communication, certain nodes are required to keep their radio on according to the union of all known schedules, and incur into higher energy consumption as a result.

8.2.1.3 Preamble Sampling-Based Protocols

In the third class of MAC protocols, targeted to low traffic loads, nodes use preamble sampling to synchronize communications. In a preamble sampling-based protocol such as the low-power listening (LPL) mechanism used in TinyOS and BMAC [29], the sender transmits a beacon to announce that it has a message to relay to a neighboring node. As depicted in Fig. 8.5, all nodes periodically turn on their radios for brief periods of time, and sample the channel looking for these beacons. If such a beacon is received, then the node keeps its radio on and waits for the actual transmission from the sender. There exist several variations of this basic mode of operations. For instance, synchronization information can be piggybacked on the beacon like in WiseMAC [14], so that the potential receivers do not have to keep their radios on, but rather can turn them off and then turn them on again at the time of communication as announced by the sender. In other protocols such as X-MAC [6], senders transmit the beacon in periodic short preambles instead of a single long one, so that receivers can acknowledge the reception of the preamble without waiting for its end. Finally, in a reversal of the original protocol, receivers may transmit the beacon to initiate transmission from the senders as it is the case in RICER [26].

8.2.1.4 Hybrid Protocols

Besides the three classes described above, there exist also hybrid MAC protocols. The objective of these protocols is to optimize performance across different traffic loads. To achieve this goal, hybrid protocols employ several techniques. For instance, they may rely on flexible MAC frame structure like IEEE 802.15.4, so that different modes of operation can be applied. In particular, the non-beacon mode of this standard protocol is basically a CSMA with collision avoidance (CSMA/CA). Instead, in the beacon-enabled mode so-called collision free period (CFPs) may be scheduled by the coordinator for specific nodes, while CSMA is still available for the rest of the time. Another solution is to blend a reservation-based protocol with a contention-based mechanism. In a hybrid protocol like ZMAC [35], nodes operate according to CSMA whenever traffic load is light, but can set up a schedule and switch to TDMA when they observe heavier traffic. Finally, in protocols such as Funneling MAC [1], nodes can operate according to a contention-based mechanism if they are further away from the sink where traffic load is low, and use reservation-based techniques if they are in the surroundings of the sink, where convergecast traffic brings about heavier loads.

8.3 Network Layer

In the past several decades, the Internet has thrived also thank to the availability of IP (the Internet Protocol) across different devices and networks. There is widespread agreement within the WSN research community that WSNs and other constrained networks will fulfill their potential, if they can seamless interoperate with the Internet. In order to enable seamless internetworking not requiring complex gateways between the Internet and constrained networks such as WSNs, it is necessary to bring the Internet network layer protocols to these novel networks. However, WSN specific features such as limited battery power and memory size make the direct implementation of network layer protocols for sensor nodes impossible. For this reason, several organizations, including the IETF, have embarked on projects to design network layer protocols that enable efficient operation of WSNs and straightforward internetworking between these and the broader Internet. As far as the IETF is concerned, the two major standard efforts are the 6LoWPAN WG with its IPv6-like protocol, and the RoLL WG with its RPL routing protocol, both depicted in Fig. 8.2. In this section, we briefly introduce 6LoWPAN and then focus our attention on the RPL routing protocol.

8.3.1 IPv6 in Low-Power Wireless Personal Area Networks

After approximately a decade of very active research in WSNs, the IETF chartered the 6LoWPAN working group to develop an IPv6-like protocol for these constrained networks using IEEE 802.15.4 at the physical and MAC layers. In order to be IPv6-compatible and work on top of IEEE 802.15.4, 6LoWPAN has to implement fragmentation, since IEEE 802.15.4 PHY frames have a maximum payload of 127 bytes, whereas IPv6 requires a 1,280-byte minimum MTU. The standard implements fragmentation using a 3-field fragmentation header. Besides a tag field for keeping track of the IPv6 packet the fragment belongs to and an offset field to keep track of its position within the IPv6 packet, 6LoWPAN also tracks the datagram size with an additional fragmentation header field as this is useful to pre-allocate a buffer of the appropriate size on resource-constrained nodes [23].

Since IEEE 802.15.4 PHY frame payload is only 127 bytes long and MAC headers use up several of them, as many as possible of the approximately 80 remaining bytes should be dedicated to carry the IPv6 payload, not the header fields. For this reason, 6LoWPAN performs stateless header compression. The adopted solution is stateless to minimize complexity on resource-constrained nodes, and is based on assigning short representations for common values in header fields while removing redundant information at the link, network, and transport layers [23]. 6LoWPAN also uses assumptions on the link layer, such as that IPv6 addresses are derived from MAC layer ones, to implement the IPv6 neighbor discovery protocol [23] for WSNs. Thanks to its support of fragmentation, header compression, and simplified neighbor discovery mechanism, 6LoWPAN is a viable solution for IPv6-based networking in WSNs. As 6LoWPAN moved through the standardization process, the need for an effort to standardize a protocol to route 6LoWPAN packets became more relevant, and the IETF RoLL working group was chartered.

8.3.2 The Routing Protocol for Low Power and Lossy Networks (RPL)

RPL is the routing protocol for low power and lossy networks under development within the IETF RoLL WG. The working group defined four different application domains for this distance vector protocol: urban environments, industrial networks, home automation, and building automation [23]. As a consequence of the selected application areas, the protocol is optimized for convergecast, supports multicast, and makes unicast communications also possible [8]. In the current version of the protocol, there is no direct support for mobility [8]. Similar to what happened within the 6LoWPAN working group, the IETF RoLL WG had to make decisions as to what would be the most important use cases and scenarios. The decision to primarily target convergecast and not to support directly mobility are rooted in the analysis of the application scenarios and the need to limit the complexity and footprint of the protocol, so that it can be adopted in novel products and applications. The correctness of these design decisions is being validated during the standardization process, and will be put to the ultimate test when the standard is released and made available to be used in real-life applications.

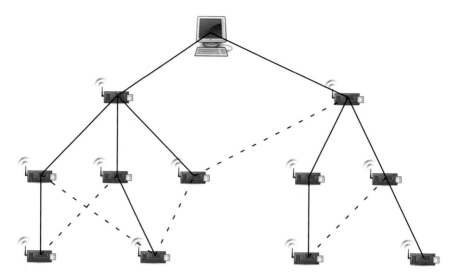

Fig. 8.6 The directed acyclic graph constructed by RPL. *Solid lines* indicate the currently selected next-hop on the route to the root, while the *dashed lines* show the other nodes in the parent set

8.3.2.1   RPL Basics

The protocol relies on an iterative process inspired by the Trickle algorithm [23] featuring one-hop DODAG (destination-oriented directed acyclic graph) Information Objects (DIOs) [8] used to propagate routing state. Instead of relying on a single node to relay packets to the root, sensors feature a parent set to achieve resilience to dynamically changing wireless links [23]. The actual next-hop neighbor is selected based on the metrics in the objective function used in the current instance of RPL [8]. RPL supports dynamic link metrics (for quality, latency, and throughput among others) in DIO messages such as ETX (estimated number of transmissions for one-hop packet transfer) [12, 23]. Figure 8.6 portrays a DAG constructed by RPL. The root also uses DODAG Confirmation Objects (DCOs) to distribute root-defined network-wide parameters [8], which are used for instance in the mechanism to repair loops. Finally, optional security mechanism is proposed [23].

8.3.2.2   Multicast and Unicast Communications

For multicast and unicast communications, RPL offers storing mode and non-storing mode [8]. In storing mode, nodes keep track of the forwarding nodes to all their descendants, so that they can re-route packets addressed to one of them, thus lowering congestion near the root. However, storing mode incurs into a larger memory footprint as nodes in the WSN must store the set of all their descendants as well as the corresponding forwarding nodes. In non-storing mode, all packets have

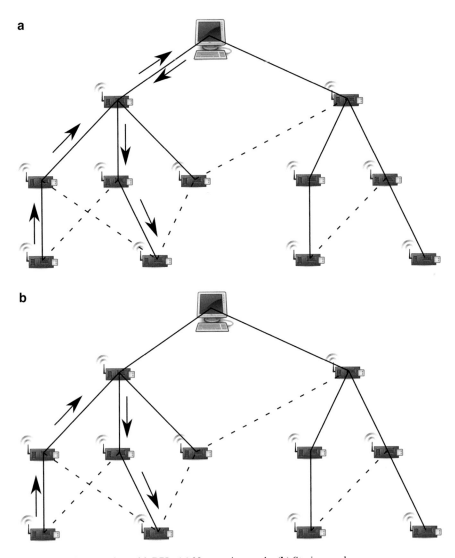

Fig. 8.7 Unicast routing with RPL. (**a**) Non-storing mode. (**b**) Storing mode

to travel all the way to the root, which then source routes them to the destination. The routes resulting from the two modes are depicted in Fig. 8.7. The advantage of non-storing mode is that nodes are relieved of the need to store information regarding their descendants. Nevertheless, non-storing mode has to surrender in terms of bandwidth and route length what it gains in terms of memory footprint. In fact, in a source routing solution, information about all hops should be included in the packet header. Since the IEEE 802.15.4 MAC frame allows approximately 80 bytes for its payload and 6LoWPAN uses several of them for its other header fields, source routing can be implemented over approximately 8 hops if uncompressed IPv6 addresses are used, and still takes away valuable payload space for upper layer (i.e., UDP) datagrams. For these reasons, we expect RPL instances to rely on storing mode more and more as memory size on resource-constrained devices slowly increases. Mixed operation with a subset of nodes using storing mode and the rest using non-storing mode is not supported [23].

8.3.3 RPL Implementations

In [22], the TinyRPL implementation of RPL and Collection Tree Protocol (CTP) [17] are evaluated by means of experiments on a testbed of 51 TelosB motes using BLIP, the Berkeley Low-Power IP stack, as the IPv6 implementation, where packets are generated every 5 and 10 s. Packet reception ratio is above 99.8% for both RPL and CTP, and between 8 and 10 control packets per hour are generated by each mote on average. The results also show that only 1.13 transmissions per hop and 1.86 end-to-end transmissions are required by TinyRPL. Unfortunately, the relevance of these results is hampered by the fact that no information is provided about the network topology, such as the average hop length of routes, on which they were collected. The authors also test the performance of bi-directional links set up by RPL, and show that a PC on the Internet sending requests to RPL motes through the edge router receives a response approximately 98% of the times. These results on round-trip packet reception ratio seem to disprove the claim in [8] discussed previously that the RPL mechanism for route construction does not select reliable bi-directional links. Finally, the authors make several suggestions for the improvement of the standard and its implementation, including a stricter definition for Trickle timer reset, trade-off solutions between storing and non-storing modes, and IPv6 fragmentation.

In [33], the authors present ContikiRPL, an implementation of RPL for Contiki OS, and discuss the results of simulations and experiments. According to the results, the ContikiRPL implementation uses approximately 3 KB ROM and 800 B RAM, which is more than an order of magnitude smaller than the 50 KB reported for it in [8]. The results for energy efficiency are encouraging as all motes maintain a duty cycle below 3% while generating 40 UDP packets per minute. However, similar to other experimental setups, the network size is limited to a few dozens motes and no information about the route hop length is provided.

8.3.4 RPL Analyses

In [36], the authors survey the research work in the area of routing as it evolved from mobile ad-hoc networks (MANETs) to WSNs. In particular, they detail flooding protocols, clustering protocols, and geographical protocols, and then the so-called self-organizing coordinate protocols. For each protocol class, the authors provide a brief overview and describe its most relevant instances. They conclude their survey by presenting RPL, an instance of gradient-based routing in the class of self-organizing coordinate systems. Thanks to its chronological approach and classification of over 40 protocols in 4 well-defined classes, this survey offers a very good insight in the research work in the area of routing for WSNs.

A detailed survey of RPL is performed in [16]. After describing the features provided by RPL and the assumed network model, the authors present the mechanisms and messages used to build the routes from the root to the sensors (used for multicast), and from the sensors to the root (used for convergecast). Unicast between sensors is implemented by using these two sets of routes in what is called dog-leg routing. The authors then present RPL mechanisms for route and loop repair, discuss several objective functions (performance metrics) used to provide QoS, and summarize security support in RPL.

RPL is experimentally evaluated on a small TelosB testbed running ContikiRPL with the Minimum Rank with Hysteresis Objective Function (MRHOF) and ETX. The results show that DODAG construction may take several minutes in a WSN of 30 nodes between 1 and 4 hops apart. The authors also measure an average power consumption of 2.2 mW during the construction of the DODAG in such a network. A packet loss ratio of 20% is observed when the RPL routes are used for multi-hop communications. The authors argue that other metrics may yield a better performance than ETX. Overall, they are satisfied with packet delays of 2.5 s in the 4-hop network. Finally, the performance of the reactive mechanism used for fault detection is also tested.

After reporting on these experiments, the authors describe other existing routing protocols and compare them to RPL. They also survey simulation and experiment results obtained by other researchers using several implementations, including ContikiRPL and TinyRPL. Finally, they point out some of the open issues in RPL, including the definition of appropriate objective functions and security mechanisms.

In [23], an overview of 6LoWPAN and RPL is presented. The survey first recaps how the research community did not consider the Internet architecture as a viable solution for communication in WSNs, thus developing many interesting, but also usually non-interoperable, ideas. It is argued that the push for the implementation of smart grids and home area networks for which WSNs are a core component prompted the adoption of the Internet architecture in WSNs. After this shift in opinion within the research community, the IETF chartered two working groups to define standards for IPv6 (6LoWPAN) and routing (RoLL) in these low-power and lossy networks, whose efforts and proposed standards are then described. The survey also briefly describes BLIP 2.0 and TinyRPL, resp. the 6LoWPAN and RPL implementations for TinyOS. According to the authors, TinyRPL with non-storing mode (the implementation with highest memory requirements) uses approximately 9 KB ROM and 300 B RAM, thus being much smaller than ContikiRPL, which uses approximately 50 KB of memory according to [8].

In [8], a critique of the current version of RPL is offered. The protocol is first described, and then analyzed by the researchers who eventually support their statements with simulation and/or experiment results, whenever feasible. It is pointed out that traffic patterns other than convergecast are also common in certain application scenarios such as building automation, but RPL has limited support for them. Furthermore, complex metrics may bring about IP fragmentation as the ICMPv6 packets carrying RPL control messages may be larger than the approximately 80 bytes allowed by IPv6 on IEEE 802.15.4. Data traffic routed using RPL in non-storing mode may also risk fragmentation, as the route needs to be incorporated in the message. With respect to storing and non-storing modes to support downward routes, it is observed that storing mode limits the route length to 64 hops if IP fragmentation is to be avoided, while the non-storing mode restricts the network size to a few dozen devices as the ones near the root need to store paths to a large subset of the WSN in their limited memory. A proposed solution to this issue when operating in storing mode consists in assigning IP addresses in the sub-tree in a hierarchical fashion as it is the case in the Internet. However, this solution limits the ability of RPL routers to change preferred parent, as all the neighbors featuring in the parent set should share a common parent for the IP address hierarchy to be maintained.

As far as bidirectional links are concerned, it is argued that selecting a preferred parent based on the link quality from it to the RPL router may not be the optimal solution as the quality of the link in the opposite direction may be very different. Furthermore, the Neighbor Unreachability Detection (NUD) mechanism proposed with RPL may be unable to detect whether the problem is indeed at this link and not farther away along the route, and to do so in a timely manner. The authors also criticize the complexity of RPL, and claim that most implementations will not be interoperable as they will have to pick a feature subset (as it is already the case for ContikiRPL [33]) in order to limit the memory footprint. They also criticize an insufficiently detailed specification, such as in the case of DAO message timing, which may lead to poor performance, and warn against Trickle performance in real-life WSNs, as its convergence is not as fast as stated by simulation results. While conceding that the RPL mechanism to support convergecast is elegant and well-understood, [8] also points out that reactive loop repair in RPL brings about potentially unacceptable delays and eventually packet losses if not all messages can be buffered at the RPL router while the loop is repaired. Furthermore, [8] also argues that mechanisms to enable unicast communication are underspecified and are likely to bring about IP fragmentation or require lots of memory for storing routes.

8.4 Application Layer

The World Wide Web is arguably one of the most successful applications enabled by the Internet. Among the several technologies making up what we call the Web, there are three fundamental ones: the HyperText Markup Language (HTML), the Hypertext Transfer Protocol (HTTP), and uniform resource identifiers (URIs). As discussed in [15, 37], HTTP implements the so-called representational state

Fig. 8.8 Internetworking via
CoAP and HTTP between the
Internet and WSNs

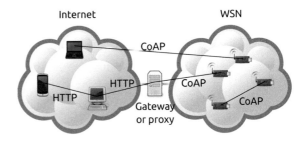

Fig. 8.9 ReSTful networking
between Internet device and
wireless sensor node

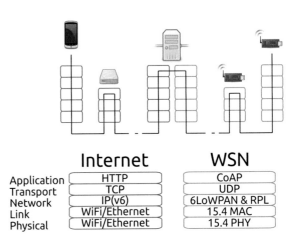

	Internet	WSN
Application	HTTP	CoAP
Transport	TCP	UDP
Network	IP(v6)	6LoWPAN & RPL
Link	WiFi/Ethernet	15.4 MAC
Physical	WiFi/Ethernet	15.4 PHY

transfer (ReST) architecture thanks to which resources (often consisting of HTML-formatted data) are accessed via their URIs. In particular, ReSTful HTTP enables interaction with remote resources identified by their URIs via four basic methods: PUT, GET, POST, and DELETE, used to create, retrieve, update, and remove resources, respectively. As the Internet of Things is slowly coming into being, researchers have started to design and analyze mechanisms to bring the powerful ReSTful paradigm to this new Internet that could potentially connect billions of devices across the world.

8.4.1 The Constrained Application Protocol (CoAP)

The Constrained Application Protocol is an application layer protocol that brings the ReST programming model of the Web to the Internet of Things and its embedded devices. Similar to HTTP, CoAP implements the four request methods of ReST; and uses similar response codes. By implementing the same ReSTful architecture as HTTP, internetworking between Web clients and CoAP-enabled WSNs as depicted in Fig. 8.8 will be streamlined. As detailed in Fig. 8.9, this will be made possible by using a simple gateway or proxy. In order to reduce complexity, CoAP relies on

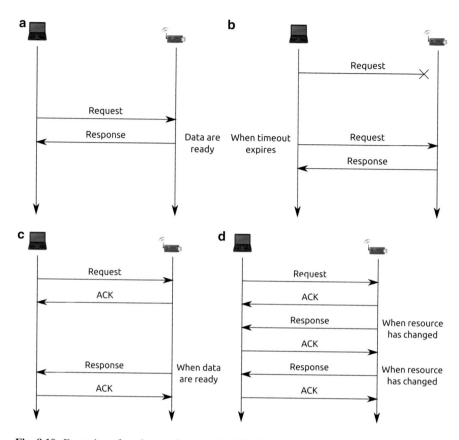

Fig. 8.10 Examples of exchanges between CoAP client and server. (**a**) Successful request-response exchange. (**b**) With packet loss and retransmission. (**c**) Using separate responses. (**d**) In observation mode

UDP instead of TCP, and defines its own simple mechanism to manage packet losses and retransmissions. Figure 8.10 portrays the sequence diagrams for communication modes provided by CoAP. CoAP supports the transfer of large payloads such as it is the case when the application or the firmware on the embedded devices need to be updated. Large payload transfer is achieved in CoAP by having multiple request-response exchanges in the so-called block mode, thus avoiding solutions involving IP fragmentation under UDP (although this is implemented by 6LoWPAN as described in Sect. 8.3), or having a stateful CoAP server. CoAP also provides a push-based mechanism, called observation, for monitoring a resource accessed via a GET request. In its GET request, the client asks the server to send a response with the current version of a resource not just once, but rather each time it changes. Finally, CoAP also addresses problems related to resource discovery for machines by defining standard resource paths on constrained devices.

In [5], the authors provide an introduction to the CoAP protocol. They first point out how standardization efforts at the network layer have brought IPv6 to WSNs (IETF 6LoWPAN), and are defining a common flexible routing protocol for these networks (IETF RoLL). Then they argue that an application layer protocol is needed that can support the growth of applications in the Internet of Things like HTTP has supported the growth of the Web. After summarizing the features of ReST, the programming model underlying the Web, the authors introduce CoAP, discussing internetworking with HTTP, and block transfer, resource observation, resource discovery, and security in the protocol. An earlier introduction to CoAP by one of the authors of [5] is provided in [30].

In [34], the authors initially summarize the major features and issues in the Internet of Things that led to the design of CoAP, and offer a brief overview thereof. The rest of the survey presents and discusses the state-of-the-art of research on CoAP in several areas: performance evaluation, comparison between CoAP-over-HTTP and SOAP-based CoAP, tools and frameworks to ease development and usage of CoAP, solutions for network configuration and service discovery, and applications to building management and the smart grid. After listing CoAP-related applications and libraries, the authors point out that support for quality of service is missing in CoAP, and argue that the dominant design for the Internet of Things (i.e., what combination of CoAP, SOAP, JSON, EXI, etc.) has not arisen yet.

8.4.2 CoAP Implementations

In [24], an implementation of CoAP for ContikiOS is presented. By relying on ContikiMAC [13], a sampling-based MAC layer protocol, it is possible to deploy an energy-efficient CoAP-enabled WSN. After summarizing MAC protocols with duty cycling and CoAP, the authors report on their implementation of CoAP for ContikiOS. This implementation provides all the protocol's major features, including block-wise transfer, resource observation, resource discovery, and separate response mechanism.

The authors run experiments on a linear 4-hop TmoteSky WSN with IEEE 802.15.4 and 6LoWPAN. The experiments show that energy-efficient operation of the CoAP-enabled WSN can be achieved simply by using an energy efficient protocol, in this case ContikiMAC, at the MAC layer without any changes to the application layer. In fact, the CoAP-enabled WSN can operate at a duty cycle around 1%, thus saving precious battery power, while latency is only lightly affected. However, the results also show that the rate of increase in latency for a CoAP exchange is higher when a duty cycling MAC protocol is used. This implies that simply relying on a duty cycling MAC for energy efficiency may result in very high latency, if the route consists of more than just a few hops as in this experimental setup.

The authors also demonstrate how ContikiMAC can help limit latency in case of block-wise transfer or 6LoWPAN fragmentation for large CoAP payloads.

In ContikiMAC, a sensor achieves this by signaling its next-hop neighbor that it will be sending a link-layer burst, i.e., a series of frames, so that the neighbor stays awake and is ready to receive them right away without going through the channel sampling stage again.

In [25], an implementation of CoAP for TinyOS and Contiki is presented, along with its application to monitor a container and its content. After introducing the application scenario and the major features of CoAP, the authors describe libcoap, their C implementation of the protocol. As such it can be readily used for the communication between the WSN and the backend, thus reducing the amount of data to be transferred over a satellite or cellular link. In order to use the CoAP library on more constrained embedded devices, it had to be stripped of some features when it was being ported to ContikiOS and TinyOS. Unfortunately, the authors do not evaluate the performance of the proposed CoAP library over a multi-hop network, but just test it over a two-node TelosB WSN. Instead, to evaluate the proposed CoAP library, the authors compare the latency and amount of data transferred over a GPRS network, which has a round-trip time similar to that of a WSN, when using CoAP and different HTTP settings, including one with bare HTTP server and client over UDP. The results show that CoAP requires 107 bytes and 1.029 s, while bare HTTP over TCP uses 885 bytes and 3.076 s.

Another comparison of CoAP and HTTP is presented in [11]. The authors first recap the adoption and adaptation of IPv6 as the standard network protocol in WSNs, and the ReSTful programming model. They then proceed to introduce two alternative stacks on top of 6LoWPAN, the IPv6 standard for embedded devices. One stack features TCP and HTTP similar to what is found in the Internet, while the other one uses UDP and CoAP. The two stacks are implemented in ContikiOS, and the authors use libcoap [25] and cURL (http://curl.haxx.se) as the respective clients to access resources on motes. In these experiments, the server and client are only one hop away from each other. The results show that CoAP exchanges consist of approximately between 10 and 20% as many bytes as HTTP ones, which is consistent with results presented in [25]. Furthermore, the authors also perform experiments using the Cooja simulator for Contiki to measure the energy consumption associated with the two stacks. The greater amount of bytes exchanged when using HTTP turns into a greater energy consumption for this protocol over CoAP. Preliminary results for transferred bytes and energy consumption were presented in [10]. The authors also perform simulations for varying request inter-arrival time and find that the energy consumption when using CoAP is not affected when requests become more frequent as it is the case for HTTP. However, the discussion of these specific results does not seem to be convincing. Finally, experiments on one-hop and two-hop routes confirm that CoAP achieves much shorter latency than HTTP.

CoAPP, an implementation of CoAP for TinyOS, is presented in [7]. Both server and client interfaces are provided in the CoAPP component with TinyOS commands in the client triggering TinyOS events in the server, and vice versa. Experiments show that 20 servers on a MEMSIC TelosB can successfully serve 90% of the 50 requests per second sent by another TelosB. Therefore, the proposed CoAP server and client implementation is deemed to be effective and scale well. The authors also implemented a library to support the encoding and decoding of XML documents into and from EXI data streams. For the EXI processing library to be used on resource-constrained microcontroller such as the Texas Instruments MSP430, the XML schemas need to be pre-processed into a set of grammars and data structures. Experiments show that the proposed EXI library is very efficient as the size of the output EXI data stream is usually around 10% of that of the original XML document containing sensor data, if a byte-aligned schema is used.

8.4.3 Internetworking Between HTTP and CoAP

A gateway and a proxy for ReSTful internetworking of CoAP and HTTP are introduced in [9]. A preliminary design of the gateway was presented in [10]. The proposed gateway consists of a Web server that presents a HTTP interface to the Web client and a CoAP client running ContikiOS that interacts with CoAP-enabled devices in the WSN. Web clients, such as browsers, are unaware of CoAP, and retrieve data from the WSN by connecting via HTTP to the Web server. The Web server then retrieves cached data from a database (Apache CouchDB [3] in this case), or uses the CoAP client to pull data from the deployed sensors.

Since a HTTP/CoAP gateway is a complex system, the authors also propose a simple HTTP-CoAP proxy. Given that both HTTP and CoAP implement the ReST programming model, the development of a HTTP-CoAP proxy is relatively straightforward. In fact, the proposed proxy offers a fully transparent protocol-agnostic resource access, so that any Web client can access a WSN using HTTP using this proxy. While the authors state that the proposed proxy does not implement resource observation, they do not list which other features have been implemented.

8.5 Discussion

Although research work on the MAC, network, and application layers of the wireless communication stack for WSN has brought about several important results, there are still several open problems in this area of research. In the rest of this section, we break down the discussion of the communication stack along the each layer and point out several open issues and problems for each one of the surveyed layers.

8.5.1 MAC Layer

At the MAC, many protocols for each one of the traffic classes have been proposed. As far as the MAC layer is concerned, an interesting question is whether the

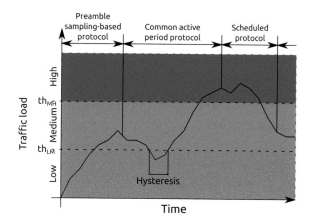

Fig. 8.11 Example of dynamic MAC protocol selection for time-varying traffic load

flexibility offered by IEEE 802.15.4 is sufficient for the diverse WSN applications looming ahead. In particular, it will be important to establish whether this standard MAC protocol can meet the requirements in terms of network lifetime for environmental monitoring applications, as well as the requirements in terms of delay and throughput of industrial applications. If research prototypes, but even more so, commercial applications demonstrate the effectiveness of IEEE 802.15.4 to achieve these different and conflicting objectives, then the standard will be widely adopted. Otherwise, there is the risk of a fragmentation in terms of the adopted MAC protocols. In that case, it may be beneficial to revisit the standard and define a more flexible solution such that nodes in a WSN, or a subnetwork thereof, can switch to the optimal operating mode within the standard for the current traffic load, be it heavy, medium, or light. Given the diversity of real-life application scenarios, we would not be surprised if they required performance levels beyond those achievable with IEEE 802.15.4.

Independent of the potential need for a more flexible standard MAC protocol, there is another important question regarding MAC protocols that has not been fully answered yet. Although MAC protocol classes have been defined for different traffic loads, there is no algorithm that can select the optimal MAC protocol given the observed traffic in a WSN. Such an algorithm should take into consideration the characteristics of the traffic. The algorithm could be employed not only for network-wide pre-deployment MAC protocol selection, but also for dynamic changes both in space and time. For instance, a preamble sampling-based protocol could be initially selected for the WSN, while a subnetwork may switch to a scheduled protocol later on to carry heavy traffic loads in a more energy-efficient manner than the original protocol. This behavior is portrayed in Fig. 8.11. Although there exist hybrid protocols and solutions that offer a simple version of this (e.g., Funneling MAC [1] with TDMA in subnetwork near the sink and CSMA in rest of the WSN), they address special situations, and do not provide a mathematical proof of their performance. This is undoubtedly a very complex problem, but advances in this area are likely to greatly benefit WSN applications by improving most performance metrics, including delay, throughput, and lifetime.

Finally, IEEE 802.15.4 or any other proposed standard MAC protocol should be extensively analyzed within the broader horizon of the communication stack. To provide the necessary background, in the following sections we introduce and discuss recent developments and advances at the network and application layers to help with this task.

8.5.2 Network Layer

After more than a decade of research work in the very important area of routing for WSNs, we are finally witnessing a strong effort by the IETF to design a standard protocol. The slightly earlier definition of an IPv6-like protocol for the WSN network layer, namely 6LoWPAN, gave the necessary boost for the chartering of IETF RoLL. Unlike application-specific protocols such as the ones in the ZigBee or Z-Wave stacks, RPL is designed to be deployed in several different scenarios, from home automation to industrial environments. Given the important role of the network layer in the Internet communication stack, it is instrumental for the growth of WSNs that the protocol be not only appropriately timed, but also flexible enough to be used in more than just one vertical market. For this reason, it is instrumental that open issues and problems be addressed swiftly during the standardization process.

An important issue is the support of traffic beyond convergecast, in particular unicast to nodes in the WSN. As we describe in more detail in Sect. 8.4, the Internet of Things is thought of as an extension of the current Internet where data collected from embedded devices such as sensor nodes can be remotely accessed via CoAP, an application layer protocol similar to HTTP. In this scenario, there is a need for requests to be sent through the WSN root over a multi-hop route to individual sensor nodes. However, the current support for this kind of communications in WSN relies on the storing or non-storing modes described above. Therefore, we argue that the performance of RPL in storing and non-storing modes be more thoroughly evaluated using CoAP traffic. Due to their shortcomings, it may be that these modes do not enable proper functioning of CoAP. In this case, the protocol should be promptly revised, and other solutions to better support unicast, such as the one suggested in [8], should be considered.

Overall, there is a need for more implementations of RPL beyond the two for TinyOS and ContikiOS, and extensive experiments on larger testbeds. In fact, several issues may arise when the protocol is tested on real-life larger deployments. First of all, there may be problems with the stability of the routes. Since the wireless medium is very noisy and its dynamics highly variable, it is unknown whether the RPL mechanism to select a forwarding node within the parent set is both optimal and stable enough in all the different application domains, from suburban homes to factory floors. In fact, it is likely that the current metrics (e.g., ETX) and objective functions (*MRHOF*) bring about frequent route updates in real-life noisy environments. As a result, timely communications in WSNs consisting of a few dozens nodes and several hops between root and leaf may become impossible. Once the performance of the protocol and, more specifically, the link metrics and objective functions is assessed, novel objective functions may be needed, that achieve (sub-) optimal but stable routes.

8.5.3   Application Layer

Although the CoAP protocol is still in the stage of Internet Draft and a standard has not been proposed yet within the IETF CoRE WG, the community around it is overall very active. This activity is demonstrated by the surveyed articles, including several ones presenting implementations of the protocol. The many analyses and implementations of the protocol also bring out several directions of further research.

First of all, we argue that the existing body of work on experimental analysis of the protocol falls short of thoroughly validating CoAP, especially when other communication stack layer are considered. For instance, while it is reassuring to know that CoAP (which uses UDP) is indeed better than HTTP (which uses TCP) in terms of transferred bytes, energy consumption, and latency, [11] this was somewhat expected as it was one of CoAP's goals from its inception. Since implementations of CoAP are already available for two of the major WSN operating systems, namely TinyOS and ContikiOS, it would be beneficial to perform an extensive experimental evaluation of this protocol. In terms of comparison of different implementations, [25] is a welcome first step, but additional effort should be dedicated to this task. Most importantly, CoAP should be evaluated on larger testbeds. Small setups such as the one consisting of 4 nodes on a path used in [24] can provide an initial validation of the protocol, but results obtained on them cannot be taken as a final proof.

Another important open issue is the necessity of energy-efficient mechanisms at the application layer. As reported in our survey of existing experiment results, it appears that the usage of an energy-efficient MAC protocol such as ContikiMAC is sufficient to greatly improve the energy efficiency of the whole communication stack [24]. While this is an important finding, we argue that it is insufficient to discard the pursuit of energy-efficient solutions at the application layer. In fact, the results in [24] were obtained for a specific traffic load, MAC protocol, and network topology. However, as we remarked in Sect. 8.2, different combinations of traffic load and MAC protocols present greatly varying behavior. For this reason, we argue that more extensive experiments with CoAP on different network topologies, or at least all traffic classes should be performed. Only the experiment results will show if the behavior observed in [24] for ContikiMAC and CoAP in presence of a relatively low traffic load extends to heavier loads and different classes of MAC protocols. In case these experiments highlight a significant performance degradation, countermeasures will have to be adopted. First of all, existing mechanisms within CoAP may be employed. For instance, separate responses could be used to counteract the increased number of retransmissions that would derive from timeouts at the client side. Alternatively, CoAP should be re-assessed and extended with novel mechanisms to support more energy-efficient operations.

Although the proposed and existing standards try to accommodate different use cases, not all application scenarios can be optimally addressed even by the most flexible standards. For instance, the proposed standards for the communication stack do not readily support in-network fusion, because the content of packets on their way from sensors to the base station cannot be inspected and modified, unless the boundaries between layers in the communication stack are broken. We argue that in most application scenarios the advantages of standardized solutions, such as interoperability of different systems, will be preferred over the positive features of customized solutions, such as a slightly reduced cost. Therefore, any solution involving in-network fusion should design, implement, and optimize it at the application layer while relying on the standard protocols at the underlying layers, rather than proposing customized cross-layer approaches.

8.6 Conclusions

In this chapter, we introduced several protocols and solutions developed to support communications in WSNs. We focused especially on the MAC, network, and application layers, due to their relevance within the communication stack. After pointing out a slow convergence of different solutions towards an Internet-like WSN communication stack featuring IEEE 802.15.4 at the physical and MAC layer, IETF 6LoWPAN and IETF RPL at the network layer, UDP at the transport layer, and IETF CoAP at the application layer, we discussed specific protocols and solutions more in detail. We observed that the research community is very active in the synthesis of many research ideas, which were proposed in the past 15 years, into well-designed standard protocols. In our discussion, we pointed out several open problems, including the selection of optimal MAC protocol for a given traffic load, objective functions that select stable routes, and the importance of energy-efficient mechanisms at layers beyond the MAC one. All these problems require more experiments to be fully modeled, and novel ideas to be solved. To conclude, we argue that, now more than ever, novel ideas solving these open problems will have the opportunity to shape standard protocols and the WSN applications of the (near) future.

References

1. Ahn G-S, Hong SG, Miluzzo E, Campbell AT, Cuomo F (2006) Funneling-MAC In: Proceedings of the 4th international conference on embedded networked sensor systems (SenSys), Boulder, p 293
2. Akyildiz IF, Su W, Sankarasubramaniam Y, Cayirci E (2002) Wireless sensor networks: a survey. Comput Netw 38(4):393–422
3. Apache Software Foundation (2011) CouchDB. Available: http://couchdb.apache.org
4. Bachir A, Dohler M, Watteyne T, Leung KK (2010) MAC essentials for wireless sensor networks. IEEE Commun Surv Tutor 12(2):222–248
5. Bormann C, Castellani AP, Shelby Z (2012) CoAP: an application protocol for billions of tiny internet nodes. IEEE Internet Comput 16(2):62–67
6. Buettner M, Yee GV, Anderson E, Han R, (2006) X-MAC: a short preamble MAC protocol for duty-cycled wireless sensor networks. In: Proceedings of the 4th international conference on embedded networked sensor systems (SenSys), Boulder, pp 307–320
7. Castellani AP, Gheda M, Bui N, Rossi M, Zorzi M (2011) Web services for the internet of things through CoAP and EXI. In: Proceedings of the IEEE international conference on communications (ICC) workshops, Kyoto, pp 1–6
8. Clausen T, Herberg U, Philipp M (2011) A critical evaluation of the IPv6 routing protocol for low power and lossy networks (RPL). In: Proceedings of the 7th IEEE international conference on wireless and mobile computing, networking and communications (WiMob), Shanghai, pp 365–372
9. Colitti W, Steenhaut K, De Caro N, Buta B, Dobrota V (2011) REST enabled wireless sensor networks for seamless integration with web applications. In: Proceedings of the 8th IEEE international conference on mobile Ad-Hoc and sensor systems (MASS), Valencia, pp 867–872
10. Colitti W, Steenhaut K, De Caro N (2011) Integrating wireless sensor networks with the web. In: Proceedings of the workshop on extending the internet to low power and lossy networks (IP+SN), Chicago,
11. Colitti W, Steenhaut K, De Caro N, Buta B, Dobrota V (2011) Evaluation of constrained application protocol for wireless sensor networks. In: Proceedings of the 18th IEEE workshop on local & metropolitan area networks (LANMAN), Chapel Hill, pp 1–6
12. Couto DSJD, Aguayo D, Bicket J, Morris R (2005) A high-throughput path metric for multi-hop wireless routing. Wirel Netw 11(4):419–434
13. Dunkels A, Mottola L, Tsiftes N, Osterlind F, Eriksson J, Finne N (2011) The announcement layer: beacon coordination for the sensornet stack. In: Wireless sensor networks, vol 6567. Springer, Berlin, pp 211–226
14. El-Hoiydi A, Decotignie J-D, Enz C, Le Roux E (2003) Poster abstract: WiseMAC, an ultra low power MAC protocol for the wiseNET wireless sensor network. In: Proceedings of the 1st international conference on embedded networked sensor systems (SenSys), Los Angeles, p 302
15. Fielding RT (2000) Architectural styles and the design of network-based software architectures. Ph.D. dissertation, University of California Irvine
16. Gaddour O, Koubâa A (2012) RPL in a nutshell: a survey. Comput Netw 56(14):3163–3178
17. Gnawali O, Fonseca R, Jamieson K, Moss D, Levis P (2009) Collection tree protocol. In: Proceedings of the 7th ACM conference on embedded networked sensor systems – SenSys'09, Berkeley, p 1
18. Hui JW, Thubert P (2011) Compression format for IPv6 datagrams over IEEE 802.15.4-based networks. Available: http://datatracker.ietf.org/doc/rfc6282
19. IEEE 802.15 Task Group 4 (TG4) (2011) IEEE Wtandard 802.15.4-2011
20. IETF CoRE (2012) Constrained RESTful environments (core). Available: http://datatracker.ietf.org/wg/core
21. INRIX Inc. (2011) INRIX traffic. Available: http://www.inrixtraffic.com

22. Ko J, Dawson-Haggerty S, Gnawali O, Culler DE, Terzis A (2011) Evaluating the performance of RPL and 6LoWPAN in TinyOS. In: Proceedings of the workshop on extending the internet to low power and lossy networks (IP+SN), Chiacgo

23. Ko J, Terzis A, Dawson-Haggerty S, Culler D, Hui J, Levis P (2011) Connecting low-power and lossy networks to the internet. IEEE Commun Mag 49(4):96–101

24. Kovatsch M, Duquennoy S, Dunkels A (2011) A low-power CoAP for contiki. In: Proceedings of the 8th IEEE international conference on mobile Ad-Hoc and sensor systems (MASS), Valencia, pp 855–860

25. Kuladinithi K, Bergmann O, Pötsch T, Becker M, Görg C (2011) Implementation of CoAP and its application in transport logistics. In: Proceedings of the workshop on extending the internet to low power and lossy networks (IP+SN), Chiacgo

26. Lin E-Y, Rabaey J, Wolisz A (2004) Power-efficient rendez-vous schemes for dense wireless sensor networks. In: Proceedings of the IEEE international conference on communications (ICC), Paris, vol 7, pp 3769–3776

27. Montenegro G, Kushalnagar N, Hui JW, Culler DE (2007) Transmission of IPv6 packets over IEEE 802.15.4 networks. Available: http://datatracker.ietf.org/doc/rfc4944

28. Pister KSJ, Doherty L (2008) TSMP: time synchronized mesh protocol. In: Proceedings of parallel and distributed computing systems (PDCS), Orlando, pp. 391–398.

29. Polastre J, Hill J, Culler D (2004) Versatile low power media access for wireless sensor networks. In: Proceedings of the 2nd international conference on embedded networked sensor systems (SenSys), Baltimore, pp 95–107

30. Shelby Z (2010) Embedded web services. IEEE Wirel Commun 17(6):52–57

31. Silva AR, Vuran MC (2010) Development of a testbed for wireless underground sensor networks. EURASIP J Wirel Commun Netw 2010:1–14

32. Tolle G, Gay D, Hong W, Polastre J, Szewczyk R, Culler D, Turner N, Tu K, Burgess S, Dawson T, Buonadonna P (2005) A macroscope in the redwoods. In: Proceedings of the 3rd international conference on embedded networked sensor systems (SenSys), San Diego, p 51

33. Tsiftes N, Eriksson J, Dunkels A (2010) Low-power wireless IPv6 routing with ContikiRPL. In: Proceedings of the 9th ACM/IEEE international conference on information processing in sensor networks (IPSN), Stockholm, pp 406–407

34. Villaverde BC, Pesch D, De Paz Alberola R, Fedor S, Boubekeur M (2012) Constrained application protocol for low power embedded networks: a survey. In: Proceedings of the 6th IEEE international conference on innovative mobile and internet services in ubiquitous computing, Palermo, pp 702–707

35. Warrier A, Aia M, Sichitiu M (2008) Z-MAC: a hybrid MAC for wireless sensor networks. IEEE/ACM Trans Netw 16(3):511–524

36. Watteyne T, Molinaro A, Richichi MG, Dohler M (2011) From MANET to IETF ROLL standardization: a paradigm shift in WSN routing protocols. IEEE Commun Surv Tutor 13(4):688–707

37. Wilde E (2007) Putting things to REST. Technical report, UC Berkeley School of Information. Available: http://datatracker.ietf.org/doc/rfc6550.

38. Winter TE, Thubert PE, Brandt A, Hui JW, Kelsey R, Levis P, Pister K, Struik R, Vasseur JP, Alexander R (2012) RPL: IPv6 routing protocol for low-power and lossy networks

39. Ye W, Heidemann J, Estrin D (2002) An energy-efficient MAC protocol for wireless sensor networks. In: Proceedings of the 21st annual joint conference of the IEEE computer and communications societies (INFOCOM), New York, no. c, pp 1567–1576

40. Yick J, Mukherjee B, Ghosal D (2008) Wireless sensor network survey. Comput Netw 52(12):2292–2330

Chapter 9
Claremont: A Solar-Powered Near-Threshold Voltage IA-32 Processor

Sriram Vangal and Shailendra Jain

9.1 Introduction to Near-Threshold Voltage (NTV) Computing

Aggressive power supply scaling into the near-threshold voltage (NTV) region holds great promise for applications with strict energy budgets. In the NTV region, the supply voltage is at or near the switching voltage (V_T) of the transistors. In this region, energy savings on the order of 5X–10X can be realized [1]. This work summarizes results from application of NTV techniques to a 32-bit Intel Architecture (IA) core in an effort to quantify and overcome the barriers that have historically relegated ultralow-voltage operation to niche markets.

The purpose of this chip is to advance NTV computing and to demonstrate the energy benefits of NTV designs, which promise better energy efficiency. Most digital designs operate at nominal voltages – about 1V today. NTV circuits operate around 400–500mV – very close to the "threshold" voltage at which transistors turn on and begin to conduct current. It is challenging to run electronics reliably at such reduced voltages. To put it simply, the difference between a "1" and a "0" in terms of electrical signal levels become very small, so a variety of noise sources can cause logic levels to be misread, leading to functional failures. The benefit, however, is that energy consumption reaches an absolute minimum in the NTV regime with a sizeable 5–10X improvement over nominal operation. The key challenge is to lock-in this excellent energy efficiency benefit at NTV while mitigating performance loss.

Enabling the processor to operate over a wide voltage range helps achieve the best possible energy efficiency while satisfying varying application performance. This work describes an IA-32 processor fabricated in 32nm CMOS technology [2], demonstrating reliable ultra-low voltage operation and energy efficient performance across the wide voltage range from 280mV to 1.2V. The research processor [3]

S. Vangal (✉) • S. Jain
Intel Labs, Intel Corporation, M/S JF2-04 2111 N.E. 25th Avenue, Hillsboro, OR 97124, USA
e-mail: sriram.r.vangal@intel.com; shailendra.jain@intel.com

P.P. Pande et al. (eds.), *Design Technologies for Green and Sustainable Computing Systems*, 229
DOI 10.1007/978-1-4614-4975-1_9, © Springer Science+Business Media New York 2013

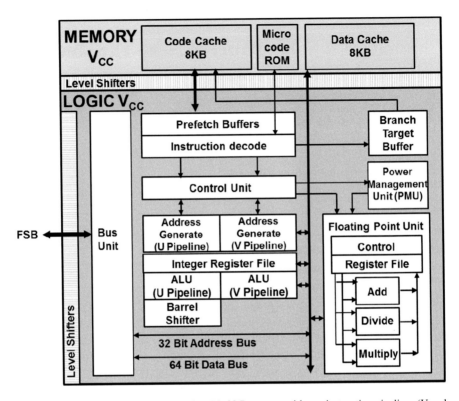

Fig. 9.1 Block diagram of Pentium™ class IA-32 Processor with *two* instruction pipelines (U and V pipelines). Processor logic and memory are on independent power planes

(Fig. 9.1) consists of a Pentium™ class IA-32 core [4] with superscalar in-order pipeline, dynamic branch prediction and 8KB of separate instruction and data caches. Core logic and memory blocks are powered by independent voltage domains to allow processor core and the memories (L1 cache + microcode ROM) to operate at their individual optimal power supplies for best overall energy efficiency. This capability allows the IA core logic to aggressively voltage scale well beyond memory Vmin limits.

9.2 NTV Circuit Design Methodology

As supply voltage approaches the threshold voltage of transistors, circuit behavior changes drastically due to an exponential increase in device delay. The presence of within Die (WID) variations results in further delay degradation. This problem becomes more prominent when the device sizes are smaller, near the process-allowed minimum width (Z_{min}), causing excessive timing push-outs and even

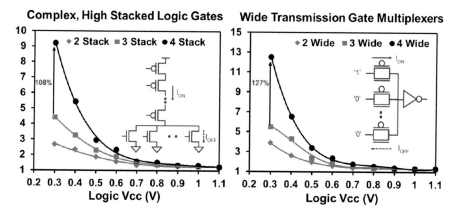

Fig. 9.2 Simulated normalized gate delays in the presence of random variations (6σ)

functional failures in case of sequential and Register File (RF) cells. This section describes low voltage design techniques used for combinational cells, sequentials, and Register File bit-cell based memory blocks.

Circuits need to be optimized for robust and reliable ultra-low voltage operation. Statistical static timing analysis (SSTA) is employed – a method which replaces the normal deterministic timing of gates and interconnects with probability distributions, and provides a distribution of possible circuit outcomes. This variation-aware SSTA study is performed on the standard cell library to eliminate the circuits which exhibit DC failures or extreme delay degradation due to reduced transistor on/off current ratios and increased sensitivity to process variations [5]. With multiple stacked devices, the drive current is significantly reduced in the NTV regime. Based on gate-level 6σ SSTA simulations (Fig. 9.2), complex logic gates with four or more stacked devices and wide transmission-gate multiplexers with four or more inputs are pruned from the library, and not used in the design, because they exhibit more than 108% and 127% delay degradation when compared to three stack gates or three-wide multiplexers respectively, at 300mV power supply.

To assist design teams with leakage power reduction while meeting performance targets, multi-threshold voltage libraries are employed with the ability to limit the use of low-voltage threshold cells. Low-voltage threshold (low V_T) cells can be good for timing, but are unfavorable for reducing power because they are very leaky. To enable reliable operation at low voltages, low V_T and high V_T devices are used selectively. All the critical timing paths are designed using low V_T devices because high V_T devices indicate 76% higher delay penalty, in the presence of variation (Fig. 9.3) at 300mV supply. Similarly, all minimum sized gates having a device width (Z_{min}) less than 2X of process-allowed minimum width are filtered from the library due to a 130% higher variation impact, when analyzed at 300mV power supply. As a result, the standard cell library was conservatively constrained, with only 40% of the total combinational cells in the library employed in the final NTV optimized design.

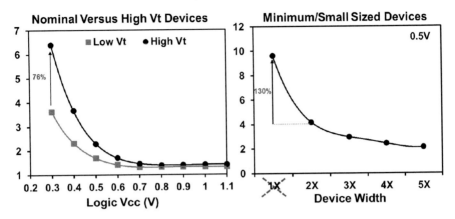

Fig. 9.3 Simulations indicate high V_T devices have 76% higher delay penalty over Low V_T flavors, while minimum width (1X) devices show 130% higher delay, at 300mV power supply

Sequential circuits and memories are more susceptible to functional failures at NTV over combinational cells, due to the need for state retention. At lower supply voltages, degradation in the transistor on/off current ratio, random and systematic process variations, affect stability of the storage nodes. Conventional transmission gate master–slave flip-flops typically have weak keepers for state nodes and larger transmission gates. During retention phase, the *on-current* of the weak keeper contends with *off-current* of the strong transmission (pass) gate affecting state node stability. Additionally, charge sharing via the pass gate between master and slave latches of the flip-flop circuit (write-back glitch between storage nodes *n1* and *n2*) can result in incorrect bit flip due to reduced noise margins at lower voltages. As a result, all sequential circuits in the NTV processor are optimized to ensure stability of state nodes in the presence of random variations. The feedback keepers are upsized to improve the state retention and are made interruptible to avoid write contention. A clocked-CMOS style flip-flop implementation (Fig. 9.4a) replaces master and slave transmission-gates in the conventional circuit topology with "pass-gate free" clocked inverters, thereby eliminating the risk of data write-back through the transmission-gate.

The processor caches employ a fully interruptible 10-transistor Register File SRAM bit cell (Fig. 9.4b) with a full transmission gate on the write bit-line (*WRBL*), which allows for contention free writes. This optimization achieves a 250mV improvement in write Vcc-min, when compared to a standard 8-transistor SRAM bit cell, at the cost of area. The bit cell is sized carefully with the help of circuit simulations to achieve 550mV retention Vcc-min. As shown in Fig. 9.4b, employment of a 10-T SRAM design can allow for operation at the lower supply voltage for optimal energy, thus making it a desirable design option for ultra-low power SRAM caches.

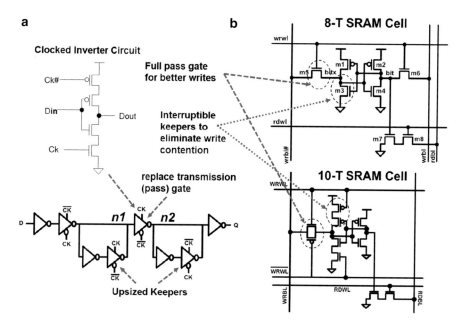

Fig. 9.4 Circuit optimizations for ultra-low voltage operation (**a**) pass-gate free low-voltage clocked-CMOS flip-flop circuit, (**b**) Original 8-T (transistor) and modified 10-T register-file interruptible cache memory bit-cell

9.3   Designing for Wide-Dynamic Range

The optimized cell library is characterized at 0.5V, 0.75V and 1.05V corners for synthesis and timing convergence. Achieving the performance targets across the entire voltage range is challenging since critical path characteristics change drastically due to non linear scaling of device delay and disproportionate scaling of device versus interconnect (wire) delay. In the absence of multi-corner, wide range design optimization tools, it is critical to identify an optimal design point such that the targeted power and performance are achieved at a given corner without a significant compromise at the other corner. Synthesis corner evaluations (Fig. 9.5) show that 0.5V, 80MHz synthesis achieves the target frequency at both 0.5V (80MHz) and 1.05V (650MHz). In comparison, it is observed that 1.05V synthesis does not sufficiently size up the device dominated data paths which become critical at lower voltages, resulting in 40% lower performance at 0.5V. Although 1.05V synthesis achieves lower leakage and better design area, the 0.5V corner was selected for final design synthesis, considering its low voltage performance benefits and promise for wide operational range.

Fig. 9.5 Optimizations for
wide range design
convergence. Design criteria
can vary widely at the 0.5V
versus1.05V corners

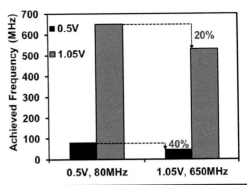

| Design | Synthesis Corner | |
Characteristics	0.5V, 80MHz	1.05V, 650MHz
Performance Range (0.5V-1.05V)	1X-8X	0.6X-6.5X
Leakage Power	1X	0.8X
Total Device Width	1X	0.78X

9.4 Experimental Results

The NTV Processor is fabricated in a 32nm CMOS process technology with nine
layers of copper interconnect [2]. The IA core is demonstrated to be operational over
the wide voltage range from 280mV to 1.2V. Figure 9.6 shows the measured total
core power and maximum operating frequency across the voltage range, measured
while running the Pentium Built-In Self-Test (BIST) in a continuous loop mode.
Starting at 1.2V and 915MHz, core voltage and performance scales down to 280mV
and 3MHz, reducing total power consumption from 737mW to merely 2mW. With
a dual-Vcc design, memories stay at its measured Vcc-min of 0.55V while allowing
logic to scale further down till 280mV.

Figure 9.7 plots the total *energy per cycle* across the wide voltage range along
with its dynamic and leakage components. Minimum energy operation is achieved at
the near-threshold voltage, with the total energy reaching minima of 170pJ/cycle at
450mV (Vcc-opt), demonstrating 4.7X improvement in energy efficiency compared
to the Vcc-max (1.2V) corner.

Figure 9.8 shows a total core power breakup across super-threshold, near-
threshold and sub-threshold regions. Contribution of logic dynamic power reduces
drastically from 81% at Vcc-max to only 4% at Vcc-min. Leakage power contribution starts increasing in the near-threshold voltage region, accounting for 42% of the
total core power at Vcc-opt. At Vcc-min point, memories continue to stay at higher
Vcc than logic, thus contributing 63% of the total core power.

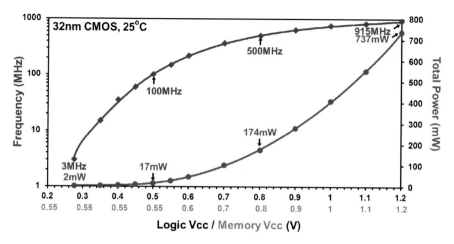

Fig. 9.6 Measured IA core power and maximum frequency of operation (Fmax) versus logic and memory power supply

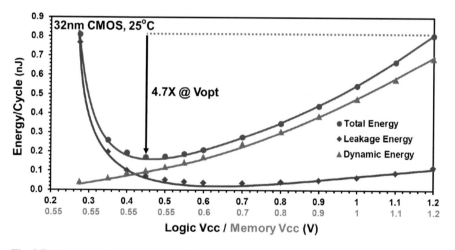

Fig. 9.7 Measured IA core energy efficiency versus logic and memory power supply. At an optimal NTV supply (V_{opt}), a 4.7X improvement in energy efficiency is observed over nominal 1.2V operation

9.5 Solar-Powered NTV Processor Demonstration

Figure 9.9 shows the packaged IA processor and the solar cell used to power the core. The 2mm² IA core contains six million transistors and uses a 951-pin flip-chip ball grid array (FCBGA) package with 168 signal pins. A custom interposer is designed to retrofit the processor into a legacy Pentium™ motherboard for silicon characterization and booting operating systems.

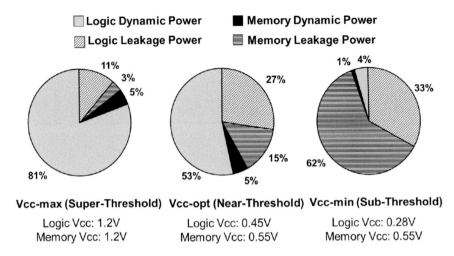

Fig. 9.8 Measured IA core power breakdown (pie-charts) from sub-threshold to super-threshold operation. Dynamic power dominates total power in the super-threshold regime while leakage power is the main contributor in the sub-threshold region, with both power components balanced in the NTV region of operation

Fig. 9.9 Packaged IA core and the solar cell used to power the core

The solar cell solution used for powering the NTV processor is shown in Fig. 9.10. A photo-voltaic cell powers an external voltage regulator module (VRM), which provides two power supply rails – a 500mV rail for the processor logic and a higher 600mV rail for the memory logic. This implementation enables

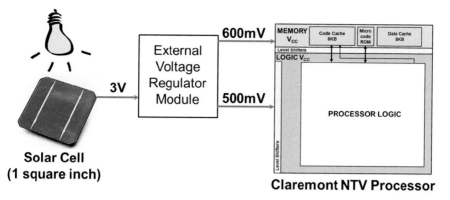

Fig. 9.10 Solar cell solution used for demonstrating the NTV processor

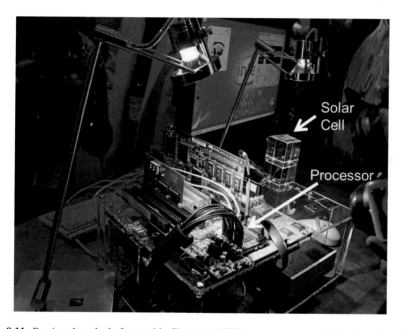

Fig. 9.11 Pentium-based platform with Claremont NTV processor powered by the solar cell. Successful windows XP™ boot is observed in the computer monitor

10–20mW of power to be harvested from the solar cell under good incandescent lighting conditions. Figure 9.11 shows a Pentium-based platform demonstration with the NTV processor and a successful Windows XP boot, with the processor core completely powered by the solar cell.

9.6   Conclusions

This case-study presented an experimental NTV IA microprocessor capable of unprecedented low-power operation. NTV technology could lead to "greener" computing, more always-on devices, longer battery lives, and energy-efficient powerful many-core processors for use in everything from handhelds to servers and even supercomputers.

Years of research went into realizing Intel's NTV IA Processor. Extreme sensitivity to power supply and transistor threshold voltage variations complicates NTV design. NTV-aware techniques had to be developed to improve design robustness for reliable operation. On-die caches were re-designed and new circuit design techniques and methods were incorporated to tolerate variations at NTV, while increasing the chip's dynamic operational range. For this test case, we selected the Pentium design, though the same techniques could be applied to any digital designs in the future.

The result is a "heat-sink free" processor core that can be placed in NTV mode at <10mW with minimum-energy and 5X better energy efficiency. The processor also provides wide dynamic operational range and can run at higher frequencies (10X) when performance is needed. The new "always-on" – yet "ultra low power state" can keep applications running and is ideal whenever compute demands are modest. Conclusions from the NTV research could lead to the integration of scalable NTV technology across a wide range of future products from mobile to high-performance computing (HPC).

NTV technology isn't just unique to processors. The concepts are promising to a wide range of digital platforms and opens up many new "use conditions", taking "always on" to a new level. For instance, this could be compelling for smart phones, tablets and other devices allowing "one" design to efficiently scale all the way, obviating the need for heterogeneous architectures. Also, these ultra-low power levels could allow energy-efficient processor architectures to expand into broader applications like embedded devices, which would include "everyday" devices such as home appliances and automobiles.

In fact, one goal of NTV research is to enable "zero power" architectures where power consumption is so low that we could power entire digital devices off solar energy, or off of the energy that surrounds us every day in the form of vibrations and ambient wireless signals. This gives us unfettered freedom so we can just leave our power cord and chargers behind. NTV research is particularly applicable to self-powered autonomous sensor networks and monitors strewn about our environment allowing computers to "see" and intelligently "react" to the world around us.

Finally, NTV research is quickly maturing and the processor is a key enabler for Extreme Scale Computing. Extreme scale means getting the most energy-efficient performance for the power spent – achieving 1000X performance at only 10X the power, or perhaps 10X performance at 1/10 the power. This could help us realize massive Exa-scale supercomputers or put trillions of computations per second in our pockets, while enabling sustainable computing along the way.

Acknowledgements The authors thank the dedicated efforts of the entire Claremont NTV processor team.

References

1. Dreslinski RG, Wieckowski M, Blaauw D, Sylvester D, Mudge T (2009) Near threshold computing: overcoming performance degradation from aggressive voltage scaling. Workshop on energy efficient design
2. Jan CH, Agostinelli M, Buehler M et al. (2009) A 32nm SoC platform technology with 2nd generation high-k/metal gate transistors optimized for ultra low power, high performance, and high density product applications. IEDM technical digest, pp 1–4
3. Jain S, Khare S, Yada S, Ambili V, Salihundam P, Ramani S, Muthukumar S, Srinivasan M, Kumar A, Gb SK, Ramanarayanan R, Erraguntla V, Howard J, Vangal S, Dighe S, Ruhl G, Aseron P, Wilson H, Borkar N, De V, Borkar S (2012) A 280mV-to-1.2V wide-operating-range IA-32 processor in 32nm CMOS. ISSCC digest of technical papers, pp 66–68
4. Schutz J (1994) A 3.3V 0.6um BiCMOS superscalar microprocessor. ISSCC digest of technical papers, pp 202–203
5. Wang A, Chandrakasan A (2004) A 180mV FFT processor using sub-threshold circuit techniques. ISSCC digest of technical papers, pp 292–592

Printed in the United States
By Bookmasters